W9-CLX-009

Lecture Notes in Mathematics

Edited by A. Dold and B. Eckmann

910

Shreeram S. Abhyankar

Weighted Expansions for Canonical Desingularization

With Foreword by U. Orbanz

Springer-Verlag
Berlin Heidelberg New York 1982

Author

Shreeram S. Abhyankar
Purdue University, Div. Math. Sci.
West Lafayette, IN 47907, USA

AMS Subject Classifications (1980): 14E15

ISBN 3-540-11195-6 Springer-Verlag Berlin Heidelberg New York
ISBN 0-387-11195-6 Springer-Verlag New York Heidelberg Berlin

This work is subject to copyright. All rights are reserved, whether the whole or part of the material is concerned, specifically those of translation, reprinting, re-use of illustrations, broadcasting, reproduction by photocopying machine or similar means, and storage in data banks. Under § 54 of the German Copyright Law where copies are made for other than private use, a fee is payable to "Verwertungsgesellschaft Wort", Munich.

© by Springer-Verlag Berlin Heidelberg 1982
Printed in Germany

Printing and binding: Beltz Offsetdruck, Hemsbach/Bergstr.
2141/3140-543210

Math

Sep.

Table of contents

FOREWORD

At the International Conference on Algebraic Geometry in La Rabida, Spain, January 1981, Prof. Abhyankar lectured on his new canonical proof of resolution of singularities in characteristic zero, giving the general idea of his procedure. Now the notes called "Weighted expansions for canonical desingularization" contain the first part of the algebraic setup to make this procedure work. In some sense these notes are disjoint from the lectures, namely they do not contain any explanation how the content is used for resolution. This foreword intends to fill this gap, at least partially.

First I describe in a very rough manner the method of resolution and some of its features. The three main ingredients of the new proof are

(1) a new refined measurement of the singularity,

(2) a canonical choice of the center to be blown up,

(3) a treatment of (1) and (2) by which the centers of blowing ups are automatically globally defined.

(1) and (2) are achieved by a new way of "expanding" an element of a regular local ring. By expansion we mean to find a certain regular system of parameters and to write the given element in terms of these parameters subject to certain (complicated) conditions. The definition of this expansion allows to take into account some regular parameters which are given in advance. In the applications, these parameters will be the ones which occured as exceptional divisors, together with their "history", i. e. the

order in which they occured. Thus (3) will be achieved by an expansion relative to given global data.

The germ for this expansion is the following procedure for plane curves. Given for example the curve defined by $f(x,z)=z^2+x^3$ at the origin, the ordinary initial form will be z^2, whereas the weighted initial form, giving weight 3/2 to z and weight 1 to x, will be z^2+x^3. Now expansion of f in the sense mentioned above consists in a choice of z and x such that z gives the multiplicity (i.e. mod x the multiplicity is unchanged) and such that among all such choices of z and x the weight that must be given to z is maximal. So this expansion comes with two numbers, the multiplicity n and the weight e, and these have the property that whenever the multiplicity is unchanged by blowing up, then after blowing up the weight will be exactly one less.

For more than two variables, the new measurement consists in an iteration of this procedure, where in each step z and x are replaced by either monomials or (weighted) homogeneous polynomials in a certain set of variables. Now for each step a third number τ has to be added, which is related to the number of "exceptional" variables used in the present step of the expansion. Then the measure of the singularity with respect to given exceptional variables will be the sequence $(n_1,e_1,\tau_1;n_2,e_2,\tau_2;\ldots)$, and the expansion to be used for resolution is one for which this sequence is maximal (in the lexicographic order). The expansion also gives the variables defining the center of blowing up, and the proof of resolution is obtained by showing that blowing up the prescribed center will improve the measure of the singularity.

We point out two major differences to Hironaka's famous proof. One is the visible change of the singularity in each step. The second is that this new proof does not use any induction on the dimension of the variety.

Even if one takes the expansion for granted, the description of resolution given above was not quite correct. The final proof will use another iteration of this procedure. After expanding one equation, one can extract some coefficients. These coefficients have to be expanded again, then the coefficients of the coefficients, etc. This leads to the notion of a web, which is not treated here.

After this rather crude description of a very complicated mechanism, we can indicate the content of the paper that follows. It contains the notation which is necessary to deal with the huge amount of information contained in the expansion. Then there is a proof for the existence of weighted initial forms in great generality, maybe more general than is needed for the purpose of resolution. Finally the existence of an expansion as indicated above is proved.

U. Orbanz

Preface

My hearty thanks to Giraud of Paris and Herrmann of Bonn whose encouragement revived my interest in resolution of singularities. I am also grateful to the Japanese gardner Hironaka for propagating sympathetic waves. But then where shall we be without the blessings of our grand master Zariski!

Indeed, Mathematics knows no national boundaries.

Our method may be termed the method of Shreedharacharya, the fifth century Indian mathematician, to whom Bhaskaracharya ascribes the device of solving quadratic equations by completing the square. The said device is given in verse number 116 of Bhaskaracharya's Bijaganita of 1150 A.D. and is thus:

चतुराहतवर्गसमै रूपैः पक्षद्वयं गुणयेत।
पूर्वाव्यक्तस्य कृतेः समरूपाणि क्षिपेत्तयोरेव इति ॥

In my youth I tried to algorithmize local resolution but had to fall back on the college algebra of rings et al for globalization. In middle age my faith in high-school algebra grew and grew to reach globalization. The lesson learnt is that when you make your local algorithms more and more precise, i.e., even more algorithmic, then they automatically globalize.

Another viewpoint. Understand desingularization of a plane curve $f(X,Y)$ better and better. Let it bloom like a lotus until it engulfs everything. Let the singletons X and Y grow into vectorial variables $X = (X_1,\dots,X_m)$ and $Y = (Y_1,\dots,Y_n)$. Let that be the petal at the core of the lotus. Now let the lotus blossom.

Or think of a beehive.

Yet another philosphical point is to understand what is a monic polynomial and thereby to enlarge that notion.

Krull and Zariski indoctrinated us with valuations and I got addicted to them. Then Hironaka taught us to rid ourselves of this addiction. But habits don't die. So now valuations have entered through the back door in their reincarnation as weights and lexicons.

This Introduction consists mostly of definitions. But from them the reader could get an idea of the proof. This is written in a pedantically precise and resultingly boring manner. I am still experimenting with notation. For me it is not easy to transcribe from the mental blackboard onto the paper!

My thanks are due to S. B. Mulay and U. Orbanz and A. M. Sathaye for stimulating discussions and help in proof-reading. Also thanks to Judy Snider for an excellent job of typing. Finally, thanks to the National Science Foundation for financial support under MCS-8002900 at Purdue University.

§1. Notation

In this paper we shall use the following notation:

Q = the set of all nonnegative rational numbers.

Z = the set of all nonnegative integers.

Z^* = the set of all subsets of Z.

$Z^{(n)}$ = the set of all n-tuples of nonnegative integers.

$[a,b] = \{n \in Z: a \leq n \leq b\}$.

§2. Semigroups

For any $u \in Q$ we put

$$\text{denom}(u) = \{0 \neq n \in Z: nu \in Z\}$$

and for any $u' \subset Q$ we put

$$\text{denom}(u') = \{0 \neq n \in Z: nu \in Z \text{ for all } u \in u'\}.$$

We note that Q is an additive abelian semigroup and Z is an (additive) subsemigroup of Q. In fact Q is a nonnegative ordered additive abelian semigroup where by a nonnegative ordered additive abelian semigroup we mean the nonnegative part of an ordered additive abelian group, i.e., the set of all nonnegative elements of an ordered additive abelian group. Likewise Z is a nonnegative ordered additive abelian semigroup. Moreover Q is divisible, but Z is not, in the following sense.

An additive abelian semigroup G is said to be divisible if for every $v \in G$ and $0 \neq n \in Z$ there exists $v^* \in G$ such that $v = nv^*$. Note that if $\bar{G}$ is a divisible ordered additive abelian group then for every $w \in \bar{G}$ and $u \in Q$ there exists a unique $w^* \in \bar{G}$ such that for every $n \in \text{denom}(u)$ we have $(nu)w = nw^*$; we define: $uw = w^*$; we observe that this notation is consistent with regarding $\bar{G}$ as a module over the ring of rational numbers; we also note that if G is the nonnegative part of $\bar{G}$ then for every $w \in G$ and $u \in Q$ we now have $uw \in G$.

§3. Strings

By a *string* we mean a system x consisting of a nonnegative integer $o(x)$, called the *length* of x, and

an element $x(c) \in$ Universe for $1 \le c \le o(x)$

whereby we call $x(c)$ the cth *component* of x. If G is a set such that $x(c) \in G$ for all c then we may call x a *G-string* or a *string on* G.

For any set G and any $o \in Z$ we put

$G(o)$ = the set of all G-strings whose length is o.

Given any additive abelian semigroup G and any $o \in Z$, we may regard $G(o)$ as an additive abelian semigroup with componentwise addition; we note that then for any $n \in Z$ and $x \in G(o)$ we have $nx \in G(o)$ given by $(nx)(c) = nx(c)$ for all $c \in [1,o]$; we also observe that if G is actually a group then so is $G(o)$; similarly, if G is a module over a ring R then in an obvious manner $G(o)$ may be regarded as a module over R. Likewise, given any divisible ordered additive abelian group G and any $o \in Z$, for any $u \in Q$ and $x \in G(o)$ we define $ux \in G(o)$ by putting $(ux)(c) = ux(c)$ for all $c \in [1,o]$.

Given any additive abelian semigroup G, for any G-string i we define

$$\text{supt}(i) = \{c \in [1,o(i)]: i(c) \neq 0\}$$

and

$$\mathrm{abs}(i) = \sum_{1 \le c \le o(i)} i(c)$$

and for any set i' of G-strings we put

$$\mathrm{abs}(i') = \{\mathrm{abs}(i) : i \in i'\}.$$

For any Q-string i we define

$$\mathrm{denom}(i) = \{0 \neq n \in Z : ni(c) \in Z \quad \text{for} \quad 1 \le c \le o(i)\}$$

and for any set i' of Q-strings we put

$$\mathrm{denom}(i') = \bigcap_{i \in i'} \mathrm{denom}(i).$$

§4. Semigroup strings with restrictions

Let o be a nonnegative integer.

For any $r \subset Z$ we put

$$\text{supt}(o,r) = [1,o] \cap r$$

and for any string D on any additive abelian semigroup we put

$$\text{supt}(o,D) = \text{supt}(o,\text{supt}(D))$$

i.e.

$$\text{supt}(o,D) = [1,o] \cap \text{supt}(D)$$

and for any $c \in Z$ we put

$$\text{supt}(o,c) = \text{supt}(o,[c,o])$$

i.e.

$$\text{supt}(o,c) = [c,o].$$

By a <u>string-subrestriction</u> we mean an object t where

$$\begin{cases} \text{either } t \subset Z \\ \text{or } t \text{ is a string on an additive abelian semigroup.} \end{cases}$$

We put

subrest(string) = the class of all string-subrestriction.

By a <u>string-restriction</u> we mean an object t where

$$\begin{cases} \text{either } t \text{ is a string-subrestriction} \\ \text{or } t \in Z \text{ .} \end{cases}$$

We put

$$\text{rest(string)} = \text{the class of all string-restrictions}$$

and we note that we have defined supt(o,t) for every $t \in$ rest(string), i.e., for every string-restriction t.

Given any string-restriction t, we put

$$\text{supt}(o,\neq t) = [1,o]\setminus \text{supt}(o,t)$$

and for any $r \subset Z$ we put

$$\text{supt}(o,t,r) = \text{supt}(o,t) \cap r$$

and for any string D on any additive abelian semigroup we put

$$\text{supt}(o,t,D) = \text{supt}(o,t,\text{supt}(D))$$

and for any $c \in Z$ we put

$$\text{supt}(o,t,c) = \text{supt}(o,t,\{c\}).$$

Thus we have defined supt(o,t,z) for any string-restrictions t and z.

Given any additive abelian semigroup G and any string-restriction t, we define

$$G(o,t) = \{i \in G(o)\colon \text{supt}(i) \subset \text{supt}(o,t)\}$$

and

$$G(o,\neq t) = \{i \in G(o)\colon \text{supt}(i) \subset \text{supt}(o,\neq t)\}$$

and for any string-restriction k we define

$$G(o,t,k) = \{i \in G(o): \text{supt}(i) \subset \text{supt}(o,t,k)\}.$$

§5. Ordered semigroup strings with restrictions

Let G be a nonnegative ordered additive abelian semigroup, let o be a nonnegative integer, and let $u \in G$.

We define

$$G(o = u) = \{i \in G(o): \mathrm{abs}(i) = u\}$$

$$G(o \geq u) = \{i \in G(o): \mathrm{abs}(i) \geq u\}$$

$$G(o > u) = \{i \in G(o): \mathrm{abs}(i) > u\}$$

$$G(o < u) = \{i \in G(o): \mathrm{abs}(i) < u\}$$

$$G(o \leq u) = \{i \in G(o): \mathrm{abs}(i) \leq u\}.$$

Given any $P \in \{=,\geq,>,<,\leq\}$ and any string-restriction t, we define

$$G(oPu,t) = G(oPu) \cap G(o,t)$$

and for any string-restriction k we define

$$G(oPu,t,k) = G(oPu) \cap G(o,t,k).$$

§6. Strings on rings

Let R be a ring (always commutative with 1). Let S be an overring of R. Let x be an S-string.

For any Z-string i with $o(i) = o(x)$ we have the element x^i in S given by

$$x^i = \prod_{1 \le c \le o(i)} x(c)^{i(c)}$$

and we put

$$x_R^i = x^i R$$

i.e.,

x_R^i = the principal R-submodule of S generated by x^i.

For any $u \in Z$ we put

$$x_R^u = \sum_{i \in Z(o(x)=u)} x_R^i$$

i.e., equivalently,

x_R^u = the R-submodule of S generated by $\{x^i : i \in Z(o(x)=u)\}$

and we note that then

$$x_R^u = (x_R^1)^u \quad \text{in case } u \neq 0$$

and we also note that

x_R^1 = the R-submodule of S generated by $x(1),\ldots,x(o(x))$.

For any $u \in Z$ and any string-restriction t we put

$$x\langle t\rangle_R^u = \sum_{i \in Z(o(x)=u,t)} x_R^i$$

i.e.,

$x\langle t\rangle_R^u$ = the R-submodule of S generated by $\{x^i: i \in Z(o(x)=u,t)\}$

and we note that then

$$x\langle t\rangle_R^u = (x\langle t\rangle_R^1)^u \quad \text{in case} \quad u \neq 0 .$$

We observe that for any $r \subset Z$ we have

$x\langle r\rangle_R^1$ = the R-submodule of S generated by $\{x(c'): c' \in [1,o(x)] \cap r\}$

and for any $c \in Z$ we have

$x\langle c\rangle_R^1$ = the R-submodule of S generated by $\{x(c'): c' \in [c,o(x)]\}$.

Here we have used, and we shall continue to use, the usual convention for sums and products which is thus.

Given any family $(G_n)_{n \in N}$ of subsets of an additive abelian semigroup G we put

$$\sum_{n \in N} G_n = \{w \in G: w = \sum_{n \in N} w_n \text{ for some } w_n \in G_n \text{ such that } w_n = 0 \text{ for all except a finite number of } n\}$$

and we note the following. If G_n is a subsemigroup of G for every $n \in N$ then $\sum_{n \in N} G_n$ is also a subsemigroup of G. If G is actually a group and G_n is a subgroup of G for every $n \in N$ then $\sum_{n \in N} G_n$ is a subgroup of G. If $G = S$ and G_n is an R-submodule of S for every $n \in N$ then $\sum_{n \in N} G_n$ is an R-submodule of S. Finally, if $G = S$ and G_n is an ideal in S for every $n \in N$ then $\sum_{n \in N} G_n$ is an ideal in S.

Given any finite disjoint sets V and W where W is nonempty, and given any $s_n \in S$ for every $n \in V$, and given any $S_n \subset S$ for every $n \in W$, we put

$$\prod_{n \in V \cup W} S_n = \Big\{ y \in S \colon y = [\prod_{n \in V} s_n] \sum_{j=1}^{m} \prod_{n \in W} z_{n,m} \text{ for some } m \in Z \text{ and some } z_{n,m} \in S_n \Big\}$$

and we note the following. If S_n is an R-submodule of S for some $n \in W$ then $\prod_{n \in V \cup W} S_n$ is also an R-submodule of S. In particular, if S_n is an ideal in S for some $n \in W$ then $\prod_{n \in V \cup W} S_n$ is also an ideal in S.

§7. Indeterminate strings

Let R be a ring.

By an <u>indeterminate string</u> over R we mean a string X such that the $X(c)$ are (algebraically) independent indeterminates over R.

Now let X be an indeterminate string over R. As usual we put

$R[X]$ = the ring of polynomials in X, [i.e., in $X(1),\ldots, X(o(X))$], with coefficients in R and exponents in Z

and we may regard $R[X]$ to be a subring of the ring $R[X]_Q$ where we put

$R[X]_Q$ = the ring of polynomials in X with coefficients in R and exponents in Q.

For any Q-string i with $o(i) = o(X)$ we have the monomial X^i, i.e., the element X^i in $R[X]_Q$ given by

$$X^i = \prod_{1\le c\le o(X)} X(c)^{i(c)}$$

and we have

$X^i_R = X^iR$

= the principal R-submodule of $R[X]_Q$ generated by X^i

and we note that:

$$X_R^i \subset R[X] \Leftrightarrow i \in Z(o(X))$$

whereas:

$$X_R^i \cap R[X] = \{0\} \Leftrightarrow i \notin Z(o(X)).$$

We observe that

$\{X^i: i \in Q(o(X))\}$ is a free R-basis of $R[X]_Q$

and so

$$R[X]_Q = \text{the R-direct sum } \sum_{i \in Q(o(X))} X_R^i .$$

For any $F \in R[X]_Q$ and $i \in Q(o(X))$, by $F[i]$ we denote the unique element in R such that

$$F = \sum_{i \in Q(o(X))} F[i]X^i .$$

For any $F \in R[X]_Q$ we define

$$\text{supt}(F) = \{i \in Q(o(X)): F[i] \neq 0\}$$

and we put

$$\text{Ord}[R,X](F) = \min \text{abs}(\text{supt}(F))$$

and

$$\text{Deg}[R,X](F) = \max \text{abs}(\text{supt}(F))$$

and we note that:

$$F \neq 0 \Leftrightarrow \mathrm{Ord}[R,X](F) \in Q \Leftrightarrow \mathrm{Deg}[R,X](F) \in Q$$

whereas:

$$F = 0 \Leftrightarrow \mathrm{Ord}[R,X](F) = \infty \Leftrightarrow \mathrm{Deg}[R,X](F) = -\infty$$

and moreover:

if $0 \neq F \in R[X]$ then $\mathrm{Ord}[R,X](F) \in Z$ and $\mathrm{Deg}[R,X](F) \in Z$.

For any $F' \subset R[X]_Q$ we put

$$\mathrm{Ord}[R,X](F') = \{\mathrm{Ord}[R,X](F): F \in F'\}$$

and

$$\mathrm{Deg}[R,X](F') = \{\mathrm{Deg}[R,X](F): F \in F'\} .$$

For any $F' \subset R[X]$ we put

$$\mathrm{Ord}[R,X]((F')) = \min \mathrm{Ord}[R,X](F')$$

and

$$\mathrm{Deg}[R,X]((F')) = \max \mathrm{Deg}[R,X](F').$$

For any $R_0 \subset R$ with $0 \in R_0$ we put

$$R_0[X]_Q = \{F \in R[X]_Q: F[i] \in R_0 \text{ for all } i \in Q(o(X))\}$$

and

$$R_0[X] = R_0[X]_Q \cap R[X].$$

Given any $u \in Q$ and $P \in \{=,\geq,>,<,\leq\}$ we define the

R-submodules $X^u_{(RP)Q}$ and $X^u_{(RP)}$ of $R[X]_Q$ and $R[X]$

respectively by putting

$$X^u_{(RP)Q} = \sum_{i \in Q(o(X)Pu)} X^i_R \quad \text{and} \quad X^u_{(RP)} = X^u_{(RP)Q} \cap R[X]$$

and for any $R_0 \subset R$ with $0 \in R_0$ we put

$$\text{Iso}(R_0,X,Pu)_Q = X^u_{(RP)Q} \cap R_0[X]_Q$$

and

$$\text{Iso}(R_0,X,Pu) = X^u_{(RP)} \cap R_0[X]$$

and we note that

$$\text{Iso}(R,X,Pu)_Q = X^u_{(RP)Q} \quad \text{and} \quad \text{Iso}(R,X,Pu) = X^u_{(RP)}$$

and we also note that

$$\text{if } P \in \{\geq,>\} \text{ then } \begin{cases} \text{Iso}(R,X,Pu)_Q \text{ is an ideal in } R[X]_Q \\ \text{and} \\ \text{Iso}(R,X,Pu) \text{ is an ideal in } R[X]. \end{cases}$$

We observe that for any $u \in Q$ we have:

$$\text{Iso}(R,X,=u)_Q = \{F \in R[X]_Q\colon \text{abs}(i) = u \text{ for all } i \in \text{supt}(F)\}$$

$$\text{Iso}(R,X,\geq u)_Q = \{F \in R[X]_Q\colon \text{Ord}[R,X](F) \geq u\}$$

$$\text{Iso}(R,X,>u)_Q = \{F \in R[X]_Q\colon \text{Ord}[R,X](F) > u\}$$

$$\mathrm{Iso}(R,X,<u)_Q = \{F \in R[X]_Q\colon \mathrm{Deg}[R,X](F) < u\}$$

$$\mathrm{Iso}(R,X,\leq u)_Q = \{F \in R[X]_Q\colon \mathrm{Deg}[R,X](F) \leq u\}$$

and we also observe that:

$$\text{if } u \in Z \text{ then } X^u_{(R\geq)} = X^u_{R[X]} = \mathrm{Iso}(R,X,\geq u).$$

For any $u \in Q$ and $P \in \{=,\geq,>,<,\leq\}$ we define the R-homomorphism

$$\mathrm{Iso}[R,X,Pu]_Q\colon R[X]_Q \to R[X]_Q$$

by putting

$$\mathrm{Iso}[R,X,Pu]_Q(F) = \sum_{i\in Q(o(X)Pu)} F[i]X^i \quad \text{for all } F \in R[X]_Q$$

and we define

$$\mathrm{Iso}[R,X,Pu]\colon R[X] \to R[X]$$

to be the R-homomorphism induced by $\mathrm{Iso}[R,X,Pu]_Q$ and we define

$$\mathrm{Iso}[R,X,Pu]^*_Q\colon R[X]_Q \to \mathrm{Iso}(R,X,Pu)_Q$$

and

$$\mathrm{Iso}[R,X,Pu]^*\colon R[X] \to \mathrm{Iso}(R,X,Pu)$$

to be the R-epimorphisms induced by $\mathrm{Iso}[R,X,Pu]_Q$.

For any $u \in Q$ we define

$$\mathrm{Info}[R,X,=u]_Q\colon \mathrm{Iso}(R,X,\geq u)_Q \to R[X]_Q$$

and

$$\mathrm{Info}[R,X,=u]\colon \mathrm{Iso}(R,X,\geq u) \to R[X]$$

to be the R-homomorphisms induced by $\mathrm{Iso}[R,X,=u]_Q$ and we define

$$\mathrm{Info}[R,X,=u]_Q^*\colon \mathrm{Iso}(R,X,\geq u)_Q \to \mathrm{Iso}(R,X,=u)_Q$$

and

$$\mathrm{Info}[R,X,=u]^*\colon \mathrm{Iso}(R,X,\geq u) \to \mathrm{Iso}(R,X,=u)$$

to be the R-epimorphisms induced by $\mathrm{Iso}[R,X,=u]_Q$ and we observe that

$$\ker(\mathrm{Info}[R,X,=u]_Q) = \ker(\mathrm{Info}[R,X,=u]_Q^*)$$

$$= \mathrm{Iso}(R,X,>u)_Q$$

and

$$\ker(\mathrm{Info}[R,X,=u]) = \ker(\mathrm{Info}[R,X,=u]^*)$$

$$= \mathrm{Iso}(R,X,>u) \ .$$

Here we have used, and we shall continue to use, the following obvious conventions for sets S_1 and S_1' and a (set-theoretic) map $g\colon S \to S'$ where $S \subset S_1$ and $S' \subset S_1'$. If S_0 is any subset of S_1 then by $g(S_0)$ we denote the subset $\{g(z)\colon z \in S_0 \cap S\}$ of S'. If $S_2 \subset S$ and $S_2' \subset S_1'$ are such that $g(S_2) \subset S_2'$ then by the map $S_2 \to S_2'$ induced by g we mean the map $g_2\colon S_2 \to S_2'$ obtained by putting $g_2(z) = g(z)$ for all $z \in S_2$. If S_3' is any subset of S_1' then by $g^{-1}(S_3')$ we denote the subset $\{z \in S: g(z) \in S_3'\}$ of S. If g is injective then by $g^{-1}: g(S) \to S$ we denote the map whereby for any $z_1 \in g(S)$ we have that $g^{-1}(z_1)$ is the unique element of S such that $g(g^{-1}(z_1)) = z_1$, i.e., such that $\{g^{-1}(z_1)\} = g^{-1}(\{z_1\})$.

§8. Indeterminate strings with restrictions

Let R be a ring, let X be an indeterminate string over R, and let t be a string-restriction.

We define the

$$\text{subrings } R[X\langle t\rangle]_Q \quad \text{and} \quad R[X\langle\neq t\rangle]_Q \quad \text{of} \quad R[X]_Q$$

by putting

$$R[X\langle t\rangle]_Q = \{f \in R[X]_Q\colon \operatorname{supt}(f) \subset Q(o(X),t)\}$$

and

$$R[X\langle\neq t\rangle]_Q = \{f \in R[X]_Q\colon \operatorname{supt}(f) \subset Q(o(X),\neq t)\}$$

and we define the

$$\text{subrings } R[X\langle t\rangle] \quad \text{and} \quad R[X\langle\neq t\rangle] \quad \text{of} \quad R[X]$$

by putting

$$R[X\langle t\rangle] = R[X\langle t\rangle]_Q \cap R[X] \quad \text{and} \quad R[X\langle\neq t\rangle] = R[X\langle\neq t\rangle]_Q \cap R[X]$$

and moreover for any $R_0 \subset R$ with $0 \in R_0$ we put

$$R_0[X\langle t\rangle]_Q = R[X\langle t\rangle]_Q \cap R_0[X]_Q \quad \text{and} \quad R_0[X\langle\neq t\rangle]_Q = R[X\langle\neq t\rangle]_Q \cap R_0[X]_Q$$

and

$$R_0[X\langle t\rangle] = R[X\langle t\rangle] \cap R_0[X] \quad \text{and} \quad R_0[X\langle\neq t\rangle] = R[X\langle\neq t\rangle] \cap R_0[X].$$

We observe that then

$$R[X\langle 1\rangle]_Q = R[X]_Q \quad \text{and} \quad R[X\langle 1\rangle] = R[X].$$

We also observe that for any $c \in [1,o(X)]$, in an obvious manner we have

$$R[X\langle\{c\}\rangle]_Q \approx R[X(c)]_Q \quad \text{and} \quad R[X\langle\{c\}\rangle] \approx R[X(c)]$$

where as usual $\approx$ stands for an isomorphism.

Given any $u \in Q$ and $P \in \{=,\geq,>,<,\leq\}$ we define the R-submodules $X\langle t\rangle^u_{(RP)Q}$ and $X\langle t\rangle^u_{(RP)}$ of $R[X\langle t\rangle]_Q$ and $R[X\langle t\rangle]$ respectively by putting

$$X\langle t\rangle^u_{(RP)Q} = \sum_{i \in Q(o(X)Pu,t)} X^i_R \quad \text{and} \quad X\langle t\rangle^u_{(RP)} = X^u_{(RP)Q} \cap R[X\langle t\rangle]$$

and for any $R_0 \subset R$ with $0 \in R_0$ we put

$$\mathrm{Iso}(R_0,X\langle t\rangle,Pu)_Q = X\langle t\rangle^u_{(RP)Q} \cap R_0[X\langle t\rangle]_Q$$

and

$$\mathrm{Iso}(R_0,X\langle t\rangle,Pu) = X\langle t\rangle^u_{(RP)} \cap R_0[X\langle t\rangle]$$

and we note that

$$\mathrm{Iso}(R,X\langle t\rangle,Pu)_Q = X\langle t\rangle^u_{(RP)Q} = \mathrm{Iso}(R,X,Pu)_Q \cap R[X\langle t\rangle]_Q$$

and

$$\mathrm{Iso}(R,X\langle t\rangle,Pu) = X\langle t\rangle^u_{(RP)} = \mathrm{Iso}(R,X,Pu)_Q \cap R[X\langle t\rangle].$$

Given any $u \in Q$ and $P \in \{=,\geq,>,<,\leq\}$ we define

$$\mathrm{Iso}[R,X\langle t\rangle,Pu]_Q\colon R[X\langle t\rangle]_Q \to R[X]_Q$$

and

$$\mathrm{Iso}[R,X\langle t\rangle,Pu]\colon R[X\langle t\rangle] \to R[X]$$

to be the R-homomorphisms induced by $\mathrm{Iso}[R,X,Pu]_Q$ and we define

$$\mathrm{Iso}[R,X\langle t\rangle,Pu]_Q^*\colon R[X\langle t\rangle]_Q \to \mathrm{Iso}(R,X\langle t\rangle,Pu)_Q$$

and

$$\mathrm{Iso}[R,X\langle t\rangle,Pu]^*\colon R[X\langle t\rangle] \to \mathrm{Iso}(R,X\langle t\rangle,Pu)$$

to be the R-epimorphisms induced by $\mathrm{Iso}[R,X,Pu]_Q$.

Given any $u \in Q$ we define

$$\mathrm{Info}[R,X\langle t\rangle,=u]_Q\colon \mathrm{Iso}(R,X\langle t\rangle,\geq u)_Q \to R[X]_Q$$

and

$$\mathrm{Info}[R,X\langle t\rangle,=u]\colon \mathrm{Iso}(R,X\langle t\rangle,\geq u) \to R[X]$$

to be the R-homomorphisms induced by $\mathrm{Iso}[R,X,=u]_Q$ and we define

$$\mathrm{Info}[R,X\langle t\rangle,=u]_Q^*\colon \mathrm{Iso}(R,X\langle t\rangle,\geq u)_Q \to \mathrm{Iso}(R,X\langle t\rangle,=u)_Q$$

and

$$\mathrm{Info}[R,X\langle t\rangle,=u]^*\colon \mathrm{Iso}(R,X\langle t\rangle,\geq u) \to \mathrm{Iso}(R,X\langle t\rangle,=u)$$

to be the R-epimorphisms induced by $Iso[R,X,=u]_Q$ and we note that

$$ker(Info[R,X\langle t\rangle,=u]_Q) = ker(Info[R,X\langle t\rangle,=u]^*_Q)$$

$$= Iso(R,X\langle t\rangle,>u)_Q$$

and

$$ker(Info[R,X\langle t\rangle,=u]) = ker(Info[R,X\langle t\rangle,=u]^*)$$

$$= Iso(R,X\langle t\rangle,>u).$$

Given any $u \in Q$ and any string-restriction k we define the

R-submodules $X\langle t,k\rangle^u_{(R=)Q}$ and $X\langle t,k\rangle^u_{(R=)}$ of $R[X\langle t\rangle]_Q$ and $R[X\langle t\rangle]$

respectively by putting

$$X\langle t,k\rangle^u_{(R=)Q} = \sum_{i\in Q(o(X)=u,t,k)} X^i_R$$

and

$$X\langle t,k\rangle^u_{(R=)} = X\langle t,k\rangle^u_{(R=)Q} \cap R[X\langle t\rangle]$$

and we define the

ideals $X\langle t,k\rangle^u_{(R\geq)Q}$ and $X\langle t,k\rangle^u_{(R\geq)}$ in $R[X\langle t\rangle]_Q$ and $R[X\langle t\rangle]$

respectively by putting

$$X\langle t,k\rangle^{u}_{(R\geq)Q} = X\langle t,k\rangle^{u}_{(R=)Q}R[X\langle t\rangle]_{Q}$$

and

$$X\langle t,k\rangle^{u}_{(R\geq)} = X\langle t,k\rangle^{u}_{(R\geq)Q} \cap R[X\langle t\rangle]$$

and for any $P \in \{=,\geq\}$, as an alternative notation, we put

$$\text{Iso}(R,X\langle t,k\rangle,Pu)_{Q} = X\langle t,k\rangle^{u}_{(RP)Q}$$

and

$$\text{Iso}(R,X\langle t,k\rangle,Pu) = X\langle t,k\rangle^{u}_{(RP)} .$$

§9. Restricted degree and order for indeterminate strings

Let R be a ring and let X be an indeterminate string over R.

Given any string-subrestriction t, we observe that $R[X]_Q$ is naturally isomorphic to $R[X\langle \neq t\rangle]_Q[X\langle t\rangle]_Q$ and this leads to the following definitions. For any $F \in R[X]_Q$ and $i \in Q(o(X),t)$, by $F\langle t\rangle[i]$ we denote the unique element in $R[X\langle \neq t\rangle]_Q$ such that

$$F = \sum_{i \in Q(o(X),t)} F\langle t\rangle[i]X^i .$$

For any $F \in R[X]_Q$ we define

$$\text{supt}(F\langle t\rangle) = \{i \in Q(o(X),t) \colon F\langle t\rangle[i] \neq 0\}$$

and we put

$$\text{Ord}[R,X](F\langle t\rangle) = \min \text{abs}(\text{supt}(F\langle t\rangle))$$

and

$$\text{Deg}[R,X](F\langle t\rangle) = \max \text{abs}(\text{supt}(F\langle t\rangle))$$

and we note that:

$$F = 0 \Leftrightarrow \text{Ord}[R,X](F\langle t\rangle) = \infty \Leftrightarrow \text{Deg}[R,X](F\langle t\rangle) = -\infty .$$

For any $F' \subset R[X]_Q$ we put

$$\text{Ord}[R,X](F'\langle t\rangle) = \{\text{Ord}[R,X](F\langle t\rangle) \colon F \in F'\}$$

and

$$\mathrm{Deg}[R,X](F'\langle t\rangle) = \{\mathrm{Deg}[R,X](F\langle t\rangle): F \in F'\}.$$

For any $F' \subset R[X]$ we put

$$\mathrm{Ord}[R,X]((F'\langle t\rangle)) = \min \mathrm{Ord}[R,X](F'\langle t\rangle)$$

and

$$\mathrm{Deg}[R,X]((F'\langle t\rangle)) = \max \mathrm{Deg}[R,X](F'\langle t\rangle).$$

Given any $c \in [1,o(X)]$, we observe that $R[X]_Q$ is naturally isomorphic to $R[X\langle\neq\{c\}\rangle]_Q[X(c)]_Q$ and this leads to the following definitions. For any $F \in R[X]_Q$ and $u \in Q$, by $F\langle c\rangle[u]$ we denote the unique element in $R[X\langle\neq\{c\}\rangle]_Q$ such that

$$F = \sum_{u\in Q} F\langle c\rangle[u]X(c)^u .$$

For any $F \in R[X]_Q$ we define

$$\mathrm{supt}(F\langle c\rangle) = \{u \in Q: F\langle c\rangle[u] \neq 0\}$$

and we put

$$\mathrm{Ord}[R,X](F\langle c\rangle) = \min \mathrm{abs}(\mathrm{supt}(F\langle c\rangle))$$

and

$$\mathrm{Deg}[R,X](F\langle c\rangle) = \max \mathrm{abs}(\mathrm{supt}(F\langle c\rangle))$$

and we note that:

$$F = 0 \Leftrightarrow \mathrm{Ord}[R,X](F\langle c\rangle) = \infty \Leftrightarrow \mathrm{Deg}[R,X](F\langle c\rangle) = -\infty \quad .$$

For any $F' \subset R[X]_Q$ we put

$$\mathrm{Ord}[R,X](F'\langle c\rangle) = \{\mathrm{Ord}[R,X](F\langle c\rangle) : F \in F'\}$$

and

$$\mathrm{Deg}[R,X](F'\langle c\rangle) = \{\mathrm{Deg}[R,X](F\langle c\rangle) : F \in F'\} \ .$$

For any $F' \subset R[X]$ we put

$$\mathrm{Ord}[R,X]((F'\langle c\rangle)) = \min \mathrm{Ord}[R,X](F'\langle c\rangle)$$

and

$$\mathrm{Deg}[R,X]((F'\langle c\rangle)) = \max \mathrm{Deg}[R,X](F'\langle c\rangle).$$

§10. Indexing strings

By an <u>indexing string</u> we mean a system ℓ consisting of

$$o(\ell) \in Z$$

$$b(\ell) \in Z \quad \text{for} \quad 1 \le b \le o(\ell)$$

and

$$T(d,b,\ell) \in Z \quad \text{for} \quad \begin{cases} 1 \le b \le o(\ell) \\ d \in Z \end{cases}$$

such that

$$o(\ell) \ne 0$$

$$b(\ell) \ne 0 \quad \text{for} \quad 1 \le b \le o(\ell) - 1$$

$$\sum_{d \in Z} T(d,b,\ell) = b(\ell) \quad \text{for} \quad 1 \le b \le o(\ell)$$

and

$$T(0,b,\ell) \begin{cases} <b(\ell) & \text{for} \quad 1 \le b \le o(\ell) - 1 \\ =b(\ell) & \text{for} \quad b = o(\ell) \end{cases} .$$

We call $o(\ell)$ the length of ℓ. For $1 \le b \le o(\ell)$ we call $b(\ell)$ the bth component of ℓ. For $1 \le b \le o(\ell)$ and $d \in Z$ we call $T(d,b,\ell)$ the $(b,d)^{th}$ <u>component</u> of ℓ.

We put

$$T(-d,b,\ell) = \sum_{0 \le n \le d} T(n,b,\ell) \qquad \text{for} \quad d \in Z.$$

Moreover we put

$$T^*[\ell] = \{s \in Z^*(o(\ell)): s(b) \subset [T(0,b,\ell)+1,b(\ell)]\}$$

and

$$T[\ell] = \text{the } Z^*\text{-string whose length is } o(\ell) \text{ and whose } b^{th} \text{ component is } [T(0,b,\ell)+1,b(\ell)] \text{ for } 1 \le b \le o(\ell)$$

and we note that

$$T[\ell] \in T^*[\ell]$$

and we put

$$T[\ell,B] = \{s \in T^*[\ell]: s(b) \neq [T(0,b,\ell)+1,b(\ell)] \text{ for some } b \in [B,o(\ell)-1]\}.$$

Finally we define

$$\text{supt}(\ell) = \{(b,c) \in Z^{(2)}: 1 \le b \le o(\ell) \text{ and } 1 \le c \le b(\ell)\}.$$

§11. Nets

By a <u>net</u> we mean a system y consisting of an indexing string $\ell(y)$, called the <u>index</u> of y, and

$$\text{an element } y(b,c) \in \text{Universe for } \begin{cases} 1 \le b \le o(\ell(y)) \\ 1 \le c \le b(\ell(y)) \end{cases}$$

whereby we call $y(b,c)$ the $(b,c)^{th}$ <u>component</u> of y. If G is a set such that $y(b,c) \in G$ for all (b,c) then we may call y a G-<u>net</u> or a <u>net on</u> G. For $1 \le b \le o(\ell(y))$, by $y(b)$ we denote the string whose length is $b(\ell(y))$ and whose c^{th} component is $y(b,c)$ for $1 \le c \le b(\ell(y))$.

For any set G and any indexing string ℓ we put

$G(\ell)$ = the set of all G-nets whose index is ℓ.

Given any additive abelian semigroup G and any indexing string ℓ, we may regard $G(\ell)$ as an additive abelian semigroup with componentwise addition; we note that then for any $n \in Z$ and $x \in G(\ell)$ we have $nx \in G(\ell)$ given by $(nx)(b,c) = nx(b,c)$ for all $(b,c) \in \text{supt}(\ell)$; we also observe that if G is actually group then so is $G(\ell)$; similarly, if G is a module over a ring R then in an obvious manner $G(\ell)$ may be regarded as a module over R. Likewise, given any divisible ordered additive abelian group G and any indexing string ℓ, for any $u \in Q$ and $x \in G(\ell)$ we define $ux \in G(\ell)$ by putting $(ux)(b,c) = ux(b,c)$ for all $(b,c) \in \text{supt}(\ell)$.

Given any additive abelian semigroup G, for any G-net j we define

$$\mathrm{supt}(j) = \{(b,c) \in \mathrm{supt}(\ell(j)) : j(b,c) \neq 0\}$$

and

$$\mathrm{abs}(j) = \sum_{\substack{1 \le b \le o(\ell(j)) \\ 1 \le c \le b(\ell(j))}} j(b,c)$$

and for any set j' of G-nets we put

$$\mathrm{abs}(j') = \{\mathrm{abs}(j) : j \in j'\} .$$

For any Q-net j we define

$$\mathrm{denom}(j) = \{0 \neq n \in Z : nj(b,c) \in Z \text{ for all } (b,c) \in \mathrm{supt}(\ell(j))\}$$

and for any set j' of Q-nets we put

$$\mathrm{denom}(j') = \bigcap_{j \in j'} \mathrm{denom}(j) .$$

§12. Semigroup nets with restrictions

Let ℓ be an indexing string.

For any $s \subset Z^{(2)}$ we put

$$\mathrm{supt}(\ell,s) = \mathrm{supt}(\ell) \cap s$$

and for any net E on any additive abelian semigroup we put

$$\mathrm{supt}(\ell,E) = \mathrm{supt}(\ell,\mathrm{supt}(E))$$

and for any Z^*-string $\hat{s}$ we put

$$\mathrm{supt}(\ell,\hat{s}) = \{(b',c') \in \mathrm{supt}(\ell) : b' \in [1,o(\hat{s})] \text{ and } c' \in \hat{s}(b')\}$$

and for any $r \subset Z$ we put

$$\mathrm{supt}(\ell,r) = \{(b',c') \in \mathrm{supt}(\ell) : b' \in r\} .$$

and for any $b \in Z$ we put

$$\mathrm{supt}(\ell,b) = \mathrm{supt}(\ell,[b,o(\ell)]) .$$

By a <u>net-subrestriction</u> we mean an object t where

$$\begin{cases} \text{either } t \subset Z^2 \\ \text{or } t \text{ is a net on an additive abelian semigroup} \\ \text{or } t \text{ is a } Z^*\text{-string} \\ \text{or } t \subset Z . \end{cases}$$

We put

$$\mathrm{subrest(net)} = \text{the class of all net-subrestrictions}$$

By a <u>net-restriction</u> we mean an object t where

$$\begin{cases} \text{either } t \text{ is a net-subrestriction} \\ \text{or } t \in Z . \end{cases}$$

We put

$$\text{rest(net)} = \text{the class of all net-restrictions}$$

and we note that we have defined $\text{supt}(\ell,t)$ for every $t \in \text{rest(net)}$, i.e., for every net-restriction t.

Given any net-restriction t, we put

$$\text{supt}(\ell,\neq t) = \text{supt}(\ell)\setminus\text{supt}(\ell,t)$$

and for any $s \subset Z^{(2)}$ we put

$$\text{supt}(\ell,t,s) = \text{supt}(\ell,t) \cap s$$

and for any net E on any additive abelian semigroup we put

$$\text{supt}(\ell,t,E) = \text{supt}(\ell,t,\text{supt}(E))$$

and for any Z^*-string $\hat{s}$ we put

$$\text{supt}(\ell,t,\hat{s}) = \text{supt}(\ell,t) \cap \text{supt}(\ell,\hat{s})$$

and for any $r \subset Z$ we put

$$\text{supt}(\ell,t,r) = \{(b',c') \in \text{supt}(\ell,t)\colon b' \in r\}$$

and for any $b \in Z$ we put

$$\text{supt}(\ell,t,b) = \{(b',c') \in \text{supt}(\ell,t)\colon b' = b\} .$$

Thus we have defined $\text{supt}(\ell,t,k)$ for any net-restrictions t and k.

Given any net-restriction t, for any $b \in Z$ and $r \subset Z$ we put

$$\text{supt}(\ell,t,b,r) = \{(b',c') \in \text{supt}(\ell,t)\colon b' = b \text{ and } c' \in r\}.$$

Given any additive abelian semigroup G and any net-restriction t, we define

$$G(\ell,t) = \{j \in G(\ell): \text{supt}(j) \subset \text{supt}(\ell,t)\}$$

and

$$G(\ell,\neq t) = \{j \in G(\ell): \text{supt}(j) \subset \text{supt}(\ell,\neq t)\}$$

and for any net-restriction k we define

$$G(\ell,t,k) = \{j \in G(\ell): \text{supt}(j) \subset \text{supt}(\ell,t,k)\}$$

and for any $b \in Z$ and $r \subset Z$ we define

$$G(\ell,t,b,r) = \{j \in G(\ell): \text{supt}(j) \subset \text{supt}(\ell,t,b,r)\}.$$

§13. Ordered semigroup nets with restrictions

Let G be a nonnegative ordered additive abelian semigroup, let ℓ be an indexing string, let $u \in G$, and let $P \in \{=,\geq,>,<,\leq\}$.

We define

$$G(\ell Pu) = \{j \in G(\ell) : \mathrm{abs}(j)Pu\} .$$

Given any net-restriction t, we define

$$G(\ell Pu,t) = G(\ell Pu) \cap G(\ell,t)$$

and for any net-restriction k we define

$$G(\ell Pu,t,k) = G(\ell Pu) \cap G(\ell,t,k)$$

and for any $b \in Z$ and $r \subset Z$ we define

$$G(\ell Pu,t,b,r) = G(\ell Pu) \cap G(\ell,t,b,r).$$

§14. Nets on rings

Let R be a ring, let S be an overring of R, and let y be an S-net.

For any Z-net j with $\ell(j) = \ell(y)$ we have the element y^j in S given by

$$y^j = \prod_{1\le b\le o(\ell(j))} y(b)^{j(b)} = \prod_{\substack{1\le b\le o(\ell(j))\\ 1\le c\le b(\ell(j))}} y(b,c)^{j(b,c)}$$

and we put

$$y_R^j = y^j R$$

i.e.,

y_R^j = the principal R-submodule of S generated by y^j.

For any $u \in Z$ we put

$$y_R^u = \sum_{j\in Z(\ell(y)=u)} y_R^j$$

i.e.,

y_R^u = the R-submodule of S generated by $\{y^j : j \in Z(\ell(y)=u)\}$

and we note that then

$$y_R^u = (y_R^1)^u \quad \text{in case} \quad u \neq 0$$

and we also note that

y_R^1 = the R-submodule of S generated by $\{y(b,c): (b,c) \in \text{supt}(\ell(y))\}$.

Given any $u \in Z$ and any net-restriction t, we put

$$y\langle t\rangle_R^u = \sum_{j \in Z(\ell(y)=u,t)} y_R^j$$

i.e.,

$y\langle t\rangle_R^u$ = the R-submodule of S generated by $\{y^j : j \in Z(\ell(y)=u,t)\}$

and we note that then

$$y\langle t\rangle_R^u = (y\langle t\rangle_R^1)^u \quad \text{in case} \quad u \neq 0 .$$

We observe that for any $s \subset Z^{(2)}$ we have

$y\langle s\rangle_R^1$ = the R-submodule of S generated by
$\{y^j: j \in \text{supt}(\ell(y)) \cap s\}$

and for any Z^*-string $\hat{s}$ we have

$y\langle\hat{s}\rangle_R^1$ = the R-submodule of S generated by
$\{y(b',c'): b' \in [1,\min(o(\ell(y)),o(\hat{s}))]$ and
$c' \in [1,b'(\ell(y))] \cap \hat{s}(b')\}$

and for any $r \subset Z$ we have

$y\langle r\rangle_R^1$ = the R-submodule of S generated by
$\{y(b',c'): b' \in [1,o(\ell(y))] \cap r$ and $c' \in [1,b(\ell(y))]\}$

and for any $b \in Z$ we have

$y\langle b\rangle_R^1$ = the R-submodule of S generated by
$\{y(b',c'): b' \in [b,o(\ell(y))]$ and $c' \in [1,b'(\ell(y))]\}$.

§15. Indeterminate nets

Let R be a ring.

By an <u>indeterminate net</u> over R we mean a net Y such that the $Y(b,c)$ are independent indeterminates over R.

We put

$R[Y]$ = the ring of polynomials in Y, [i.e., in $Y(b,c)_{1\le b\le o(\ell(Y)),1\le c\le b(\ell(Y))}$], with coefficients in R and exponents in Z

and we may regard $R[Y]$ to be a subring of the ring $R[Y]_Q$ where we put

$R[Y]_Q$ = the ring of polynomials in Y with coefficients in R and exponents in Q.

For any $b \in [1,o(\ell(Y))]$ and any $i \in Q(b(\ell(Y)))$ we have the monomial

$$Y(b)^i = \prod_{1\le c\le o(i)} Y(b,c)^{i(c)} \in R[Y]_Q .$$

For any $j \in Q(\ell(Y))$ we have the monomial

$$Y^j = \prod_{1\le b\le o(\ell(j))} Y(b)^{j(b)} = \prod_{\substack{1\le b\le o(\ell(j))\\ 1\le c\le b(\ell(j))}} Y(b,c)^{j(b,c)} \in R[Y]_Q$$

and we have

$$Y_R^j = Y^j R$$

$$= \text{the principal } R\text{-submodule of } R[Y]_Q \text{ generated by } Y^j$$

and we note that:

$$Y_R^j \subset R[Y] \Leftrightarrow j \in Z(\ell(Y))$$

whereas:

$$Y_R^j \cap R[Y] = \{0\} \Leftrightarrow j \notin Z(\ell(Y)).$$

We observe that

$$\{Y^j : j \in Q(\ell(Y))\} \text{ is a free } R\text{-basis of } R[Y]_Q$$

and so

$$R[Y]_Q = \text{the } R\text{-direct sum } \sum_{j \in Q(\ell(Y))} Y_R^j .$$

For any $F \in R[Y]_Q$ and $j \in Q(\ell(Y))$, by $F[j]$ we denote the unique element in R such that

$$F = \sum_{j \in Q(\ell(Y))} F[j]Y^j .$$

For any $F \in R[Y]_Q$ we define

$$\text{supt}(F) = \{j \in Q(\ell(Y)) : F[j] \neq 0\}$$

and we put

$$\mathrm{Ord}[R,Y](F) = \min \mathrm{abs}(\mathrm{supt}(F))$$

and

$$\mathrm{Deg}[R,Y](F) = \max \mathrm{abs}(\mathrm{supt}(F))$$

and we note that:

$$F \neq 0 \Leftrightarrow \mathrm{Ord}[R,Y](F) \in Q \Leftrightarrow \mathrm{Deg}[R,Y](F) \in Q$$

whereas:

$$F = 0 \Leftrightarrow \mathrm{Ord}[R,Y](F) = \infty \Leftrightarrow \mathrm{Deg}[R,X](F) = -\infty$$

and moreover:

if $0 \neq F \in R[Y]$ then $\mathrm{Ord}[R,Y](F) \in Z$ and $\mathrm{Deg}[R,Y](F) \in Z$.

For any $F' \subset R[Y]_Q$ we put

$$\mathrm{Ord}[R,Y](F') = \{\mathrm{Ord}[R,Y](F) : F \in F'\}$$

and

$$\mathrm{Deg}[R,Y](F') = \{\mathrm{Deg}[R,Y](F) : F \in F'\}.$$

For any $F' \subset R[Y]$ we put

$$\mathrm{Ord}[R,Y]((F')) = \min \mathrm{Ord}\ [R,Y](F')$$

and

$$\mathrm{Deg}[R,Y]((F')) = \max \mathrm{Deg}[R,Y](F').$$

For any $R_0 \subset R$ with $0 \in R_0$ we put

$$R_0[Y]_Q = \{F \in R[Y]_Q : F[j] \in R_0 \text{ for all } j \in Q(\ell(Y))\}$$

and

$$R_0[Y] = R_0[Y]_Q \cap R[Y] .$$

Given any $u \in Q$ and $P \in \{=,\geq,>,<,\leq\}$ we define the R-submodules $Y^u_{(RP)Q}$ and $Y^u_{(RP)}$ of $R[Y]_Q$ and $R[Y]$ respectively by putting

$$Y^u_{(RP)Q} = \sum_{j \in Q(\ell(Y)Pu)} Y^i_R \quad \text{and} \quad Y^u_{(RP)} = Y^u_{(RP)Q} \cap R[Y]$$

and for any $R_0 \subset R$ with $0 \in R_0$ we put

$$\mathrm{Iso}(R_0,Y,Pu)_Q = Y^u_{(RP)Q} \cap R_0[Y]_Q$$

and

$$\mathrm{Iso}(R_0,Y,Pu) = Y^u_{(RP)} \cap R_0[Y]$$

and we note that

$$\mathrm{Iso}(R,Y,Pu)_Q = Y^u_{(RP)Q} \quad \text{and} \quad \mathrm{Iso}(R,Y,Pu) = Y^u_{(RP)}$$

and we also note that

$$\text{if } P \in \{\geq,>\} \text{ then } \begin{cases} \mathrm{Iso}(R,Y,Pu)_Q \text{ is an ideal in } R[Y]_Q \\ \text{and} \\ \mathrm{Iso}(R,Y,Pu) \text{ is an ideal in } R[Y]. \end{cases}$$

We observe that for any $u \in Q$ we have:

$$\mathrm{Iso}(R,Y,=u)_Q = \{F \in R[Y]_Q : \mathrm{abs}(j) = u \text{ for all } j \in \mathrm{supt}(F)\}$$

$$\mathrm{Iso}(R,Y,\geq u)_Q = \{F \in R[Y]_Q : \mathrm{Ord}[R,Y](F) \geq u\}$$

$$\mathrm{Iso}(R,Y,>u)_Q = \{F \in R[Y]_Q : \mathrm{Ord}[R,Y](F) > u\}$$

$$\mathrm{Iso}(R,Y,<u)_Q = \{F \in R[Y]_Q : \mathrm{Deg}[R,Y](F) < u\}$$

$$\mathrm{Iso}(R,Y,\leq u)_Q = \{F \in R[Y]_Q : \mathrm{Deg}[R,Y](F) \leq u\}$$

and we also observe that:

$$\text{if } u \in Z \text{ then } Y^u_{(R\geq)} = Y^u_{R[Y]} = \mathrm{Iso}(R,Y,\geq u).$$

For any $u \in Q$ and $P \in \{=,\geq,>,<,\leq\}$ we define the R-homomorphism

$$\mathrm{Iso}[R,Y,Pu]_Q : R[Y]_Q \to R[Y]_Q$$

by putting

$$\mathrm{Iso}[R,Y,Pu]_Q(F) = \sum_{j \in Q(\ell(Y)Pu)} F[j]Y^j \quad \text{for all} \quad F \in R[Y]_Q$$

and we define

$$\mathrm{Iso}[R,Y,Pu]: R[Y] \to R[Y]$$

to be the R-homomorphism induced by $\mathrm{Iso}[R,Y,Pu]_Q$ and we define

$$\mathrm{Iso}[R,Y,Pu]^*_Q: R[Y]_Q \to \mathrm{Iso}(R,Y,Pu)_Q$$

and

$$\mathrm{Iso}[R,Y,Pu]^*: R[Y] \to \mathrm{Iso}(R,Y,Pu)$$

to be the R-epimorphisms induced by $\mathrm{Iso}[R,Y,Pu]_Q$.

For any $u \in Q$ we define

$$\mathrm{Info}[R,Y,=u]_Q: \mathrm{Iso}(R,Y,\geq u)_Q \to R[Y]_Q$$

and

$$\mathrm{Info}[R,Y,=u]: \mathrm{Iso}(R,Y,\geq u) \to R[Y]$$

to be the R-epimorphisms induced by $\mathrm{Iso}[R,Y,=u]_Q$ and we define

$$\mathrm{Info}[R,Y,=u]^*_Q: \mathrm{Iso}(R,Y,\geq u)_Q \to \mathrm{Iso}(R,Y,=u)_Q$$

and

$$\mathrm{Info}[R,Y,=u]^*: \mathrm{Iso}(R,Y,\geq u) \to \mathrm{Iso}(R,Y,=u)$$

to be the R-epimorphisms induced by $\mathrm{Iso}[R,Y,=u]_Q$ and we observe that

$$\ker(\mathrm{Info}[R,Y,=u]_Q) = \ker(\mathrm{Info}[R,Y,=u]^*_Q) = \mathrm{Iso}(R,Y,>u)_Q$$

and

$$\ker(\mathrm{Info}[R,Y,=u]) = \ker(\mathrm{Info}[R,Y,=u]^*) = \mathrm{Iso}(R,Y,>u).$$

§16. Indeterminate nets with restrictions

Let R be a ring, let Y be an indeterminate net over R, and let t be a net-restriction.

We define the

subrings $R[Y\langle t\rangle]_Q$ and $R[Y\langle \neq t\rangle]_Q$ of $R[Y]_Q$

by putting

$$R[Y\langle t\rangle]_Q = \{f \in R[Y]_Q : \operatorname{supt}(f) \subset Q(\ell(Y),t)\}$$

and

$$R[Y\langle \neq t\rangle]_Q = \{f \in R[Y]_Q : \operatorname{supt}(f) \subset Q(\ell(Y),\neq t)\}$$

and we define the

subrings $R[Y\langle t\rangle]$ and $R[Y\langle \neq t\rangle]$ of $R[Y]$

by putting

$$R[Y\langle t\rangle] = R[Y\langle t\rangle]_Q \cap R[Y] \quad \text{and} \quad R[Y\langle \neq t\rangle] = R[Y\langle \neq t\rangle]_Q \cap R[Y],$$

and moreover for any $R_0 \subset R$ with $0 \in R_0$ we put

$$R_0[Y\langle t\rangle]_Q = R[Y\langle t\rangle]_Q \cap R_0[Y]_Q \quad \text{and} \quad R_0[Y\langle \neq t\rangle]_Q = R[Y\langle \neq t\rangle]_Q \cap R_0[Y]_Q$$

and

$$R_0[Y\langle t\rangle] = R[Y\langle t\rangle] \cap R_0[Y] \quad \text{and} \quad R_0[Y\langle \neq t\rangle] = R[Y\langle \neq t\rangle] \cap R_0[Y] .$$

We observe that then

$$R[Y\langle 1\rangle]_Q = R[Y]_Q \quad \text{and} \quad R[Y\langle 1\rangle] = R[Y] \ .$$

We also observe that for any $b \in [1,o(\ell(Y))]$, in an obvious manner we have

$$R[Y\langle\{b\}\rangle]_Q \approx R[Y(b)]_Q \quad \text{and} \quad R[Y\langle\{b\}\rangle] \approx R[Y(b)]$$

and moreover for any $c \in [1,b(\ell(Y))]$, in an obvious manner we have

$$R[Y\langle\{(b,c)\}\rangle]_Q \approx R[Y(b,c)]_Q \quad \text{and} \quad R[Y\langle\{(b,c)\}\rangle] \approx R[Y(b,c)].$$

Given any $u \in Q$ and $P \in \{=,\geq,>,<,\leq\}$ we define the R-submodules $Y\langle t\rangle^u_{(RP)Q}$ and $Y\langle t\rangle^u_{(RP)}$ of $R[Y\langle t\rangle]_Q$ and $R[Y\langle t\rangle]$ respectively by putting

$$Y\langle t\rangle^u_{(RP)Q} = \sum_{j\in Q(\ell(Y)Pu,t)} Y^j_R \quad \text{and} \quad Y\langle t\rangle^u_{(RP)} = Y\langle t\rangle^u_{(RP)Q} \cap R[Y\langle t\rangle]$$

and for any $R_0 \subset R$ with $0 \in R_0$ we put

$$\mathrm{Iso}(R_0,Y\langle t\rangle,Pu)_Q = Y\langle t\rangle^u_{(RP)Q} \cap R_0[Y\langle t\rangle]_Q$$

and

$$\mathrm{Iso}(R_0,Y\langle t\rangle,Pu) = Y\langle t\rangle^u_{(RP)} \cap R_0[Y\langle t\rangle]$$

and we note that

$$\mathrm{Iso}(R,Y\langle t\rangle,Pu)_Q = Y\langle t\rangle^u_{(RP)Q} = \mathrm{Iso}(R,Y,Pu)_Q \cap R[Y\langle t\rangle]_Q$$

and

$$\mathrm{Iso}(R,Y\langle t\rangle,Pu) = Y\langle t\rangle^u_{(RP)} = \mathrm{Iso}(R,Y,Pu)_Q \cap R[Y\langle t\rangle].$$

Given any $u \in Q$ and $P \in \{=,\geq,>,<,\leq\}$ we define

$$\mathrm{Iso}[R,Y\langle t\rangle,Pu]_Q\colon R[Y\langle t\rangle]_Q \to R[Y]_Q$$

and

$$\mathrm{Iso}[R,Y\langle t\rangle,Pu]\colon R[Y\langle t\rangle] \to R[Y]$$

to be the R-homomorphisms induced by $\mathrm{Iso}[R,Y,Pu]_Q$ and we define

$$\mathrm{Iso}[R,Y\langle t\rangle,Pu]^*_Q\colon R[Y\langle t\rangle]_Q \to \mathrm{Iso}(R,Y\langle t\rangle,Pu)_Q$$

and

$$\mathrm{Iso}[R,Y\langle t\rangle,Pu]^*\colon R[Y\langle t\rangle] \to \mathrm{Iso}(R,Y\langle t\rangle,Pu)$$

to be the R-epimorphisms induced by $\mathrm{Iso}[R,Y,Pu]_Q$.

Given any $u \in Q$ we define

$$\mathrm{Info}[R,Y\langle t\rangle,=u]_Q\colon \mathrm{Iso}(R,Y\langle t\rangle,\geq u)_Q \to R[Y]_Q$$

and

$$\text{Info}[R,Y\langle t\rangle,=u]\colon \text{Iso}(R,Y\langle t\rangle,\geq u) \to R[Y]$$

to be the R-homomorphisms induced by $\text{Iso}[R,Y,=u]_Q$ and we define

$$\text{Info}[R,Y\langle t\rangle,=u]_Q^*\colon \text{Iso}(R,Y\langle t\rangle,\geq u)_Q \to \text{Iso}(R,Y\langle t\rangle,=u)_Q$$

and

$$\text{Info}[R,Y\langle t\rangle,=u]^*\colon \text{Iso}(R,Y\langle t\rangle,\geq u) \to \text{Iso}(R,Y\langle t\rangle,=u)$$

to be the R-epimorphism induced by $\text{Iso}[R,Y,=u]_Q$ and we note that

$$\begin{aligned}\ker(\text{Info}[R,Y\langle t\rangle,=u]_Q) &= \ker(\text{Info}[R,Y\langle t\rangle,=u]_Q^*)\\ &= \text{Iso}(R,Y\langle t\rangle,>u)_Q\end{aligned}$$

and

$$\begin{aligned}\ker(\text{Info}[R,Y\langle t\rangle,=u]) &= \ker(\text{Info}[R,Y\langle t\rangle,=u]^*)\\ &= \text{Iso}(R,Y\langle t\rangle,>u).\end{aligned}$$

Given any $u \in Q$ and any net-restriction k we define the R-submodules $Y\langle t,k\rangle^u_{(R=)Q}$ and $Y\langle t,k\rangle^u_{(R=)}$ of $R[Y\langle t\rangle]_Q$ and $R[Y\langle t\rangle]$ respectively by putting

$$Y\langle t,k\rangle^u_{(R=)Q} = \sum_{j\in Q(\ell(Y)=u,t,k)} Y^j_R$$

and

$$Y\langle t,k\rangle^{u}_{(R=)} = Y\langle t,k\rangle^{u}_{(R=)Q} \cap R[Y\langle t\rangle]$$

and we define the

ideals $Y\langle t,k\rangle^{u}_{(R\geq)Q}$ and $Y\langle t,k\rangle^{u}_{(R\geq)}$ in $R[Y\langle t\rangle]_Q$ and $R[Y\langle t\rangle]$

respectively by putting

$$Y\langle t,k\rangle^{u}_{(R\geq)Q} = Y\langle t,k\rangle^{u}_{(R=)Q}R[Y\langle t\rangle]_Q$$

and

$$Y\langle t,k\rangle^{u}_{(R\geq)} = Y\langle t,k\rangle^{u}_{(R\geq)Q} \cap R[Y\langle t\rangle]$$

and for any $P \in \{=,\geq\}$, as an alternative notation, we put

$$\mathrm{Iso}(R,Y\langle t,k\rangle,Pu)_Q = Y\langle t,k\rangle^{u}_{(RP)Q}$$

and

$$\mathrm{Iso}(R,Y\langle t,k\rangle,Pu) = Y\langle t,k\rangle^{u}_{(RP)} .$$

Given any $u \in Q$ and given any $b \in Z$ and $r \subset Z$, we define the

R-submodules $Y\langle t,b,r\rangle^{u}_{(R=)Q}$ and $Y\langle t,b,r\rangle^{u}_{(R=)}$ of $R[Y\langle t\rangle]_Q$ and $R[Y\langle t\rangle]$

respectively by putting

$$Y\langle t,b,r\rangle^{u}_{(R=)Q} = \sum_{j\in Q(\ell(Y)=u,t,b,r)} Y^{j}_{R}$$

and

$$Y\langle t,b,r\rangle^{u}_{(R=)} = Y\langle t,b,r\rangle^{u}_{(R=)Q} \cap R[Y\langle t\rangle]$$

and we define the

ideals $Y\langle t,b,r\rangle^{u}_{(R\geq)Q}$ and $Y\langle t,b,r\rangle^{u}_{(R\geq)}$ in $R[Y\langle t\rangle]_Q$ and $R[Y\langle t\rangle]$

respectively by putting

$$Y\langle t,b,r\rangle^{u}_{(R\geq)Q} = Y\langle t,b,r\rangle^{u}_{(R=)Q}\, R[Y\langle t\rangle]_Q$$

and

$$Y\langle t,b,r\rangle^{u}_{(R\geq)} = Y\langle t,b,r\rangle^{u}_{(R\geq)Q} \cap R[Y\langle t\rangle]$$

and for any $P \in \{=,\geq\}$, as an alternative notation, we put

$$\mathrm{Iso}(R,Y\langle t,b,r\rangle,Pu)_Q = Y\langle t,b,r\rangle^{u}_{(RP)Q}$$

and

$$\mathrm{Iso}(R,Y\langle t,b,r\rangle,Pu) = Y\langle t,b,r\rangle^{u}_{(RP)} .$$

Given any $u \in Q$ and $P \in \{=,\geq\}$ and given any $b \in [1,o(\ell(Y))]$ and $s \in Z^*(o(\ell(Y)))$, we put

$$Y\langle t,b,s\rangle^{u}_{(RP)Q} = Y\langle t,b,s(b)\rangle^{u}_{(RP)Q}$$

and

$$Y\langle t,b,s\rangle^{u}_{(RP)} = Y\langle t,b,s(b)\rangle^{u}_{(RP)}$$

and as an alternative notation we put

$$\mathrm{Iso}(R,Y\langle t,b,s\rangle,Pu)_Q = Y\langle t,b,s\rangle^u_{(RP)Q}$$

and

$$\mathrm{Iso}(R,Y\langle t,b,s\rangle,Pu) = Y\langle t,b,s\rangle^u_{(RP)} \quad .$$

§17. Restricted degree and order for indeterminate nets

Let R be a ring and let Y be an indeterminate net over R.

Given any net-subrestriction t, we observe that $R[Y]_Q$ is naturally isomorphic to $R[Y\langle\neq t\rangle]_Q[Y\langle t\rangle]_Q$ and this leads to the following definitions. For any $F \in R[Y]_Q$ and $j \in Q(\ell(Y),t)$, by $F\langle t\rangle[j]$ we denote the unique element in $R[Y\langle\neq t\rangle]_Q$ such that

$$F = \sum_{j \in Q(\ell(Y),t)} F\langle t\rangle[j]Y^j .$$

For any $F \in R[Y]_Q$ we define

$$\mathrm{supt}(F\langle t\rangle) = \{j \in Q(\ell(Y),t): F\langle t\rangle[j] \neq 0\}$$

and we put

$$\mathrm{Ord}[R,Y](F\langle t\rangle) = \min \mathrm{abs}(\mathrm{supt}(F\langle t\rangle))$$

and

$$\mathrm{Deg}[R,Y](F\langle t\rangle) = \max \mathrm{abs}(\mathrm{supt}(F\langle t\rangle))$$

and we note that:

$$F = 0 \Leftrightarrow \mathrm{Ord}[R,Y](F\langle t\rangle) = \infty \Leftrightarrow \mathrm{Deg}[R,Y](F\langle t\rangle) = -\infty .$$

For any $F' \subset R[Y]_Q$ we put

$$\mathrm{Ord}[R,Y](F'\langle t\rangle) = \{\mathrm{Ord}[R,Y](F\langle t\rangle): F \in F'\}$$

and

$$\mathrm{Deg}[R,Y](F'\langle t\rangle) = \{\mathrm{Deg}[R,Y](F\langle t\rangle): F \in F'\} .$$

For any $F' \subset R[Y]$ we put

$$\mathrm{Ord}[R,Y]((F'\langle t\rangle)) = \min \mathrm{Ord}[R,Y](F'\langle t\rangle)$$

and

$$\mathrm{Deg}[R,Y]((F'\langle t\rangle)) = \max \mathrm{Deg}[R,Y](F'\langle t\rangle).$$

Given any $b \in [1,o(\ell(Y))]$, we observe that $R[Y]_Q$ is naturally isomorphic to $R[Y\langle\neq\{b\}\rangle]_Q[Y(b)]_Q$ and this leads to the following definitions. For any $F \in R[Y]_Q$ and $i \in Q(b(\ell(Y)))$, by $F\langle b\rangle[i]$ we denote the unique element in $R[Y\langle\neq\{b\}\rangle]_Q$ such that

$$F = \sum_{i\in Q(b(\ell(Y)))} F\langle b\rangle[i]Y(b)^i .$$

For any $F \in R[Y]_Q$ we define

$$\mathrm{supt}(F\langle b\rangle) = \{i \in Q(b(\ell(Y))): F\langle b\rangle[i] \neq 0\}$$

and we put

$$\mathrm{Ord}[R,Y](F\langle b\rangle) = \min \mathrm{abs}(\mathrm{supt}(F\langle b\rangle))$$

and

$$\mathrm{Deg}[R,Y](F\langle b\rangle) = \max \mathrm{abs}(\mathrm{supt}(F\langle b\rangle))$$

and we note that:

$$F = 0 \Leftrightarrow \mathrm{Ord}[R,Y](F\langle b\rangle) = \infty \Leftrightarrow \mathrm{Deg}[R,Y](F\langle b\rangle) = -\infty .$$

For any $F' \subset R[Y]_Q$ we put

$$\mathrm{Ord}[R,Y](F'\langle b\rangle) = \{\mathrm{Ord}[R,Y](F\langle b\rangle) : F \in F'\}$$

and

$$\mathrm{Deg}[R,Y](F'\langle b\rangle) = \{\mathrm{Deg}[R,Y](F\langle b\rangle) : F \in F'\}.$$

For any $F' \subset R[Y]$ we put

$$\mathrm{Ord}[R,Y]((F'\langle b\rangle)) = \min \mathrm{Ord}[R,Y](F'\langle b\rangle)$$

and

$$\mathrm{Deg}[R,Y]((F'\langle b\rangle)) = \max \mathrm{Deg}[R,Y](F'\langle b\rangle).$$

Given any $b \in [1,o(\ell(Y))]$ and $c \in [1,b(\ell(Y))]$, we observe that $R[Y]_Q$ is naturally isomorphic to $R[Y\langle\neq\{(b,c)\}]_Q[Y(b,c)]_Q$ and this leads to the following definitions. For any $F \in R[Y]_Q$ and $u \in Q$, by $F\langle b,c\rangle[u]$ we denote the unique element in $R[Y\langle\neq\{(b,c)\}\rangle]_Q$ such that

$$F = \sum_{u\in Q} F\langle b,c\rangle[u]Y(b,c)^u.$$

For any $F \in R[Y]_Q$ we define

$$\mathrm{supt}(F\langle b,c\rangle) = \{u \in Q : F\langle b,c\rangle[u] \neq 0\}$$

and we put

$$\mathrm{Ord}[R,Y](F\langle b,c\rangle) = \min \mathrm{supt}(F\langle b,c\rangle)$$

and

$$\mathrm{Deg}[R,Y](F\langle b,c\rangle) = \max \mathrm{supt}(F\langle b,c\rangle)$$

and we note that:

$$F = 0 \Leftrightarrow \mathrm{Ord}[R,Y](F\langle b,c\rangle) = \infty \Leftrightarrow \mathrm{Deg}[R,Y](F\langle b,c\rangle) = -\infty$$

For any $F' \subset R[Y]_Q$ we put

$$\mathrm{Ord}[R,Y](F'\langle b,c\rangle) = \{\mathrm{Ord}[R,Y](F\langle b,c\rangle) : F \in F'\}$$

and

$$\mathrm{Deg}[R,Y](F'\langle b,c\rangle) = \{\mathrm{Deg}[R,Y](F\langle b,c\rangle) : F \in F'\}.$$

For any $F' \subset R[Y]$ we put

$$\mathrm{Ord}[R,Y]((F'\langle b,c\rangle)) = \min \mathrm{Ord}[R,Y](F'\langle b,c\rangle)$$

and

$$\mathrm{Deg}[R,Y]((F'\langle b,c\rangle)) = \max \mathrm{Deg}[R,Y](F'\langle b,c\rangle).$$

§18. Prechips

By a <u>prechip</u> we mean a system e consisting of an indexing string $\ell(e)$ called the <u>index</u> of e and

a nonnegative rational number $e(B,b,c)$ for $\begin{cases} 1 \le B \le o(\ell(e)) \\ 1 \le b \le o(\ell(e)) \\ 0 \le c \le b(\ell(e)) \end{cases}$

called the $(B,b,c)^{th}$ <u>component</u> of e, such that

$$e(B,b,c) = \begin{cases} 0 & \text{if } B = o(\ell(e)) \text{ and } c \ne 0 \\ 0 & \text{if } b < B \\ 0 & \text{if } b > B \text{ and } c = 0 \\ 0 & \text{if } c > T(-1,b,\ell(e)). \end{cases}$$

We put

$$e[B] = e(B,B,0) \quad \text{for} \quad 1 \le B \le o(\ell(e))$$

and

$$e[[B]] = \sum_{\substack{1 \le b \le o(\ell(e)) \\ 0 \le c \le b(\ell(e))}} e(B,b,c) \quad \text{for} \quad 1 \le B \le o(\ell(e)).$$

For $1 \le B \le o(\ell(e))$, by $e(B)$ we denote the Q-net whose index is $\ell(e)$ and whose

$(b,c)^{th}$ component is $e(B,b,c)$ for $\begin{cases} 1 \le b \le o(\ell(e)) \\ 1 \le c \le b(\ell(e)). \end{cases}$

For $1 \le B \le o(\ell(e))$ and $1 \le b \le o(\ell(e))$, by $e(B,b)$ we denote the Q-string whose length is $b(\ell(e))$ and whose c^{th} component is $e(B,b,c)$ for $1 \le c \le b(\ell(e))$.

We define

$$\text{denom}(e) = \{0 \ne n \in Z\colon ne(B,b,c) \in Z \text{ for } 1 \le B \le o(\ell(e)) \text{ and } 1 \le b \le o(\ell(e)) \text{ and } 0 \le c \le b(\ell(e))\}$$

and for any set e' of prechips we put

$$\text{denom}(e') = \bigcap_{e \in e'} \text{denom}(e).$$

For any $u \in Q$, by ue we denote the prechip with $\ell(ue) = \ell(e)$ such that $(ue)(B,b,c) = ue(B,b,c)$ for $1 \le B \le o(\ell(e))$ and $1 \le b \le o(\ell(e))$ and $0 \le c \le b(\ell(e))$. Likewise, given any indexing string ℓ, we may regard the set of all prechips whose index is ℓ as an additive abelian semigroup with componentwise addition.

§19. Isobars for prechips and Premonic polynomials

Let R be a ring. Let Y be an indeterminate net over R. Let e be a prechip with $\ell(e) = \ell(Y)$. Let B' be an integer with $1 \le B' \le o(\ell(e)) - 1$.

For any $u \in Q$ and $P \in \{=,\ge\}$ we define the

R-submodule $Y\langle B',B\rangle^{u}_{(R,eP)Q}$ of $R[Y\langle B'\rangle]_Q$ for $B \in [B',o(\ell(e))-1]$

by first putting

$$Y\langle B',B\rangle^{u}_{(R,eP)Q} = Y\langle B',B\rangle^{u}_{(RP)Q} \qquad \text{in case } B' \le B = o(\ell(e)) - 1$$

and then, by decreasing induction on B, putting

$$Y\langle B',B\rangle^{u}_{(R,eP)Q}$$

$$= \sum_{\substack{v\in Q \text{ and } w\in Q \\ \text{with } v+w=u}} Y\langle B',B\rangle^{v}_{(RP)Q} Y^{we(B+1)} Y\langle B',B+1\rangle^{we[B+1]}_{(R,eP)Q}$$
$$\text{in case } B' \le B \le o(\ell(e)) - 2$$

where we recall that $we(B+1)$ denotes the net whose index is $\ell(e)$ and whose $(b,c)^{\text{th}}$ component is $we(B+1,b,c)$ for $1 \le b \le o(\ell(e))$ and $1 \le c \le b(\ell(e))$, and where we take note that because of the defining conditions of a prechip we have

$$Y^{we(B+1)} \in R[Y\langle B'+1\rangle]_Q \subset R[Y\langle B'\rangle]_Q \quad .$$

Given any $u \in Q$ and $B \in [B',o(\ell(e)) - 1]$, for any $P \in \{=,\geq\}$ we define the

R-submodule $Y\langle B',B\rangle^{u}_{(R,eP)}$ of $R[Y\langle B'\rangle]$

by putting

$$Y\langle B',B\rangle^{u}_{(R,eP)} = Y\langle B',B\rangle^{u}_{(R,eP)Q} \cap R[Y\langle B'\rangle]$$

and for any $R_0 \subset R$ with $0 \in R_0$ we put

$$\mathrm{Iso}(R_0,Y\langle B',B\rangle,ePu)_Q = Y\langle B',B\rangle^{u}_{(R,eP)Q} \cap R_0[Y]_Q$$

and

$$\mathrm{Iso}(R_0,Y\langle B',B\rangle,ePu) = Y\langle B',B\rangle^{u}_{(R,eP)} \cap R_0[Y]$$

and we observe that

$\mathrm{Iso}(R,Y\langle B',B\rangle,e \geq u)_Q$ is an ideal in $R[Y\langle B'\rangle]_Q$

and

$\mathrm{Iso}(R,Y\langle B',B\rangle,e \geq u)$ is an ideal in $R[Y\langle B'\rangle]$.

Given any $u \in Q$ and $B \in [B',o(\ell(e)) - 1]$, for any $P \in \{=,\geq\}$ we define the

R-submodule $Y\langle B',B\rangle^{u}_{((R,eP))Q}$ of $R[Y\langle B'\rangle]_Q$

by putting

$$Y\langle B',B\rangle^{u}_{((R,eP))Q} = Y^{ue(B)} Y\langle B',B\rangle^{ue[B]}_{(R,eP)Q}$$

and we define the

R-submodule $Y\langle B',B\rangle^{u}_{((R,eP))}$ of $R[Y\langle B'\rangle]$

by putting

$$Y\langle B',B\rangle^{u}_{((R,eP))} = Y\langle B',B\rangle^{u}_{((R,eP))Q} \cap R[Y\langle B'\rangle]$$

and for any $R_0 \subset R$ with $0 \in R_0$ we put

$$\mathrm{Iso}((R_0, Y\langle B',B\rangle, ePu))_Q = Y\langle B',B\rangle^{u}_{((R,eP))Q} \cap R_0[Y]_Q$$

and

$$\mathrm{Iso}((R_0, Y\langle B',B\rangle, ePu)) = Y\langle B',B\rangle^{u}_{((R,eP))} \cap R_0[Y]$$

and we observe that

$\mathrm{Iso}((R, Y\langle B',B\rangle, e \geq u))_Q$ is an ideal in $R[Y\langle B'\rangle]_Q$

and

$\mathrm{Iso}((R, Y\langle B',B\rangle, e \geq u))$ is an ideal in $R[Y\langle B'\rangle]$.

We note that for any $u \in Q$, $B \in [B', o(\ell(e)) - 2]$, and $P \in \{=,\geq\}$ we have

$$Y\langle B',B\rangle^{u}_{(R,eP)Q} = \sum_{\substack{v\in Q \text{ and } w\in Q \\ \text{with } v+w=u}} Y\langle B',B\rangle^{v}_{(RP)Q}\, Y\langle B',B+1\rangle^{w}_{((R,eP))Q} .$$

For any $u \in Q$ and for any $s \in Z^*(o(\ell(e)))$ and $P \in \{=,\geq\}$ we define the

R-submodule $Y\langle B',B,s\rangle^u_{(R,eP)Q}$ of $R[Y\langle B'\rangle]_Q$ for $B \in [B',o(\ell(e))-1]$

by first putting

$$Y\langle B',B,s\rangle^u_{(R,eP)Q} = Y\langle B',B,s\rangle^u_{(RP)Q} \text{ in case } B' \leq B = o(\ell(e)) - 1$$

and then, by decreasing induction on B, putting

$$Y\langle B',B,s\rangle^u_{(R,eP)Q} = Y\langle B',B,s\rangle^u_{(RP)Q}$$

$$+ \sum_{\substack{v\in Q \text{ and } 0\neq w\in Q \\ \text{with } v+w=u}} Y\langle B',B\rangle^v_{(RP)Q} Y^{we(B+1)} Y\langle B',B+1,s\rangle^{we[B+1]}_{(R,eP)Q}$$

$$\text{in case } B' \leq B \leq o(\ell(e)) - 2 \; .$$

Given any $u \in Q$ and given any $B \in [B',o(\ell(e)) - 1]$ and $s \in Z^*(o(\ell(e)))$, for any $P \in \{=,\geq\}$ we define the

R-submodule $Y\langle B',B,s\rangle^u_{(R,eP)}$ of $R[Y\langle B'\rangle]$

by putting

$$Y\langle B',B,s\rangle^u_{(R,eP)} = Y\langle B',B,s\rangle^u_{(R,eP)Q} \cap R[Y\langle B'\rangle]$$

and for any $R_0 \subset R$ with $0 \in R_0$ we put

$$\mathrm{Iso}(R_0, Y\langle B', B, s\rangle, ePu)_Q = Y\langle B', B, s\rangle^u_{(R,eP)Q} \cap R_0[Y]_Q$$

and

$$\mathrm{Iso}(R_0, Y\langle B', B, s\rangle, ePu) = Y\langle B', B, s\rangle^u_{(R,eP)} \cap R_0[Y]$$

and we observe that

$$\mathrm{Iso}(R, Y\langle B', B, s\rangle, e \geq u)_Q \text{ is an ideal in } R[Y\langle B'\rangle]_Q$$

and

$$\mathrm{Iso}(R, Y\langle B', B, s\rangle, e \geq u) \text{ is an ideal in } R[Y\langle B'\rangle].$$

Given any $u \in Q$ and given any $B \in [B', o(\ell(e)) - 1]$ and $s \in Z^*(o(\ell(e)))$, for any $P \in \{=, \geq\}$ we define the

$$\text{R-submodule } Y\langle B', B, s\rangle^u_{((R,eP))Q} \text{ of } R[Y\langle B'\rangle]_Q$$

by putting

$$Y\langle B', B, s\rangle^u_{((R,eP))Q} = Y^{ue(B)} Y\langle B', B, s\rangle^{ue[B]}_{(R,eP)Q}$$

and we define the

$$\text{R-submodule } Y\langle B', B, s\rangle^u_{((R,eP))} \text{ of } R[Y\langle B'\rangle]$$

by putting

$$Y\langle B', B, s\rangle^u_{((R,eP))} = Y\langle B', B, s\rangle^u_{((R,eP))Q} \cap R[Y\langle B'\rangle]$$

and for any $R_0 \subset R$ with $0 \in R_0$ we put

$$\mathrm{Iso}((R_0, Y\langle B',B,s\rangle, ePu))_Q = Y\langle B',B,s\rangle^u_{((R,eP))Q} \cap R_0[Y]_Q$$

and

$$\mathrm{Iso}((R_0, Y\langle B',B,s\rangle, ePu)) = Y\langle B',B,s\rangle^u_{((R,eP))} \cap R_0[Y]$$

and we observe that

$\mathrm{Iso}((R, Y\langle B',B,s\rangle, e \geq u))_Q$ is an ideal in $R[Y\langle B'\rangle]_Q$

and

$\mathrm{Iso}((R, Y\langle B',B,s\rangle, e \geq u))$ is an ideal in $R[Y\langle B'\rangle]$.

We note that for any $u \in Q$, $s \in Z^*(o(\ell(e)))$, $B \in [B', o(\ell(e)) - 2]$, and $P \in \{=, \geq\}$ we have

$$Y\langle B',B,s\rangle^u_{(R,eP)Q} = Y\langle B',B,s\rangle^u_{(RP)Q}$$

$$+ \sum_{\substack{v \in Q \text{ and } 0 \neq w \in Q \\ \text{with } v+w=u}} Y\langle B',B\rangle^v_{(RP)Q} Y\langle B',B+1,s\rangle^w_{((R,eP))Q}.$$

Given any $u \in Q$ and given any $B \in [B', o(\ell(e)) - 1]$ and $s' \subset Z^*(o(\ell(e)))$, for any $P \in \{=, \geq\}$ we define

$$Y\langle B',B,s'\rangle^u_{(R,eP)Q} = \bigcup_{s \in s'} Y\langle B',B,s\rangle^u_{(R,eP)Q}$$

$$Y\langle B',B,s'\rangle^u_{(R,eP)} = \bigcup_{s \in s'} Y\langle B',B,s\rangle^u_{(R,eP)}$$

$$Y\langle B',B,s'\rangle^{u}_{((R,eP))Q} = \bigcup_{s\in s'} Y\langle B',B,s\rangle^{u}_{((R,eP))Q}$$

$$Y\langle B',B,s'\rangle^{u}_{((R,eP))} = \bigcup_{s\in s'} Y\langle B',B,s\rangle^{u}_{((R,eP))}$$

and for any $R_0 \subset R$ with $0 \in R_0$ we put

$$\mathrm{Iso}(R_0,Y\langle B',B,s'\rangle,ePu)_Q = Y\langle B',B,s'\rangle^{u}_{(R,eP)Q} \cap R_0[Y]_Q$$

$$\mathrm{Iso}(R_0,Y\langle B',B,s'\rangle,ePu) = Y\langle B',B,s'\rangle^{u}_{(R,eP)} \cap R_0[Y]$$

$$\mathrm{Iso}((R_0,Y\langle B',B,s'\rangle,ePu))_Q = Y\langle B',B,s'\rangle^{u}_{((R,eP))Q} \cap R_0[Y]_Q$$

$$\mathrm{Iso}((R_0,Y\langle B',B,s'\rangle,ePu)) = Y\langle B',B,s'\rangle^{u}_{((R,eP))} \cap R_0[Y].$$

For any $u \in Q$ and $B \in [B',o(\ell(e)) - 1]$ and $R_0 \subset R$ with $0 \in R_0$ we define

$$\mathrm{Nonpremon}(R_0,Y\langle B',B\rangle,e=u)$$

$$= \mathrm{Iso}(R_0,Y\langle B',B,T[\ell(e),B]\rangle,e=u)$$

and

$$\mathrm{Premon}(R_0,Y\langle B',B\rangle,e=u)$$

$$= \mathrm{Iso}(R_0,Y\langle B',B,T[\ell(e)]\rangle,e=u)\backslash \mathrm{Nonpremon}(R_0,Y\langle B',B\rangle,e=u)$$

and we define

$$\mathrm{Nonpremon}((R_0, Y\langle B', B\rangle, e = u))$$

$$= \mathrm{Iso}((R_0, Y\langle B', B, T[\ell(e), B]\rangle, e = u))$$

and

$$\mathrm{Premon}((R_0, Y\langle B', B\rangle, e = u))$$

$$= \mathrm{Iso}((R_0, Y\langle B', B, T[\ell(e)]\rangle, e=u)) \setminus \mathrm{Nonpremon}((R_0, Y\langle B', B\rangle, e=u)).$$

§20. Substitutions

Let R be a ring. Let Y be given where

$$\begin{cases} \text{either } Y \text{ is an indeterminate string over } R \\ \text{or } Y \text{ is an indeterminate net over } R. \end{cases}$$

Given any $\bar{Y}$ where

$$\begin{cases} \bar{Y} \in R[Y](o(Y)) \text{ in case } Y \text{ is a string} \\ \text{whereas } \bar{Y} \in R[Y](\ell(Y)) \text{ in case } Y \text{ is a net} \end{cases}$$

we define the R-algebra-homomorphism

$$\mathrm{Sub}[R,Y=\bar{Y}]: R[Y] \to R[Y]$$

by putting

$$\mathrm{Sub}[R,Y=\bar{Y}](F) = \sum_{j\in \mathrm{supt}(F)} F[j]\bar{Y}^j \quad \text{for all } F \in R[Y]$$

and for any $R_0 \subset R$ with $0 \in R_0$ we put

$$R_0[\bar{Y}] = \mathrm{Sub}[R,Y=\bar{Y}](R_0[Y])$$

and we define

$$\mathrm{Sub}[R,Y=\bar{Y}]^*: R[Y] \to R[\bar{Y}]$$

to be the R-algebra-epimorphism induced by $\mathrm{Sub}[R,Y=\bar{Y}]$.

Given any y where

$$\begin{cases} y \in R(o(Y)) \text{ in case } Y \text{ is a string} \\ \text{whereas } y \in R(\ell(Y)) \text{ in case } Y \text{ is a net} \end{cases}$$

we define

$$\text{sub}[R, Y = y]: R[Y] \to R$$

to be the R-algebra-epimorphism induced by $\text{Sub}[R, Y = y]$ and we note that then

$$\text{sub}[R, Y = y] = \text{Sub}[R, Y = y]^* .$$

By a <u>pseudomorphism</u> we mean a (set-theoretic) map $g: R \to R'$ such that R' is a ring and $g(0) = 0$ and $g(1) = 1$.

Given any pseudomorphism $g: R \to R'$ and given any Y' where

$$\begin{cases} Y' \in R'[Y](o(Y)) & \text{in case } Y \text{ is a string} \\ \text{whereas } Y' \in R'[Y](\ell(Y)) & \text{in case } Y \text{ is a net} \end{cases}$$

we define the pseudomorphism

$$\text{Sub}[g, Y = Y']: R[Y] \to R'[Y]$$

by putting

$$\text{Sub}[g, Y = Y'](F) = \sum_{j \in \text{supt}(F)} g(F[j]) Y'^{j} \quad \text{for all } F \in R[Y]$$

and we define

$$\text{Sub}[g, Y = Y']^*: R[Y] \to g(R)[Y']$$

to be the surjective map induced by $\text{Sub}[g, Y = Y']$ and we note that: if g is a ring-homomorphism then $\text{Sub}[g, Y = Y']$ and

$Sub[g, Y = Y']^*$ are ring-homomorphisms and $Sub[g, Y = Y']$ is a g-algebra-homomorphism.

Given any pseudomorphism $g: R \to R'$ and given any y' where

$$\begin{cases} y' \in R'(o(Y)) & \text{in case } Y \text{ is a string} \\ \text{whereas } y' \in R'(\ell(Y)) & \text{in case } Y \text{ is a net} \end{cases}$$

we define

$$sub[g, Y = y']: R[Y] \to R'$$

to be the pseudomorphism induced by $Sub[g, Y = y']$ and we define

$$sub[g, Y = y']^*: R[Y] \to g(R)[y']$$

to be the surjective map induced by $sub[g, Y = y']$ and we note that then

$$sub[g, Y = y']^* = Sub[g, Y = y']^*$$

and we observe that: if g is a ring-homomorphism then $sub[g, Y = y']$ and $sub[g, Y = y']^*$ are ring-homomorphisms and $sub[g, Y = y']$ is a g-algebra-homomorphism.

Given any pseudomorphism $g: R \to R'$ we define the pseudomorphism

$$Sub[g, Y]_Q: R[Y]_Q \to R'[Y]_Q$$

by putting

$$Sub[g, Y]_Q(F) = \sum_{j \in supt(F)} g(F[j])Y^j \quad \text{for all } F \in R[Y]_Q$$

and we define

$$Sub[g,Y]: R[Y] \to R'[Y]$$

to be the pseudomorphism induced by $Sub[g,Y]_Q$ and we define

$$Sub[g,Y]^*_Q: R[Y]_Q \to g(R)[Y]_Q$$

and

$$Sub[g,Y]^*: R[Y] \to g(R)[Y]$$

to be the surjective maps induced by $Sub[g,Y]_Q$ and we note that then

$$Sub[g,Y] = Sub[g,Y=Y] \quad \text{and} \quad Sub[g,Y]^* = Sub[g,Y=Y]^*$$

and we observe that: if g is a ring-homomorphism then $Sub[g,Y]_Q$ and $Sub[g,Y]^*_Q$ are ring-homomorphisms and $Sub[g,Y]_Q$ is a g-algebra-homomorphism.

We define the R-algebra-homomorphism

$$Sub[R,Y=0]_Q: R[Y]_Q \to R[Y]_Q$$

by putting

$$Sub[R,Y=0]_Q(F) = \begin{cases} \sum_{j\in Q(o(Y)=0)} F[j] & \text{in case } Y \text{ is a string} \\ \sum_{j\in Q(\ell(Y)=0)} F[j] & \text{in case } Y \text{ is a net} \end{cases}$$

and we define

$$Sub[R,Y=0]: R[Y] \to R[Y]$$

to be the R-algebra-homomorphism induced by $\mathrm{Sub}[R,Y=0]_Q$ and we define

$$\mathrm{Sub}[R,Y=0]_Q^*: R[Y]_Q \to R$$

$$\mathrm{Sub}[R,Y=0]^*: R[Y] \to R$$

and

$$\mathrm{sub}[R,Y=0]: R[Y] \to R$$

to be the R-algebra-epimorphisms induced by $\mathrm{Sub}[R,Y=0]_Q$ and we note that

$$\mathrm{sub}[R,Y=0] = \mathrm{Sub}[R,Y=0]^*.$$

Given any pseudomorphism $g: R \to R'$, we define the pseudomorphism

$$\mathrm{Sub}[g,Y=0]_Q: R[Y]_Q \to R'[Y]_Q$$

by putting

$$\mathrm{Sub}[g,Y=0]_Q(F) = \begin{cases} \sum_{j\in Q(o(Y)=0)} g(F[j]) & \text{in case } Y \text{ is a string} \\ \sum_{j\in Q(\ell(Y)=0)} g(F[j]) & \text{in case } Y \text{ is a net} \end{cases}$$

and we define

$$\mathrm{Sub}[g,Y=0]: R[Y] \to R'[Y]$$

and

$$\mathrm{sub}[g,Y=0]: R[Y] \to R'$$

to be the pseudomorphism induced by $\mathrm{Sub}[g, Y = 0]_Q$ and we define

$$\mathrm{Sub}[g, Y = 0]_Q^*: R[Y]_Q \to g(R)$$

$$\mathrm{Sub}[g, Y = 0]^*: R[Y] \to g(R)$$

and

$$\mathrm{sub}[g, Y = 0]^*: R[Y] \to g(R)$$

to be the surjective maps induced by $\mathrm{Sub}[g, Y = 0]_Q$ and we note that

$$\mathrm{sub}[g, Y = 0]^* = \mathrm{Sub}[g, Y = 0]^*$$

and we also observe that: if g is a ring-homomorphism then the above defined six maps are ring-homomorphisms and out of them the first three are g-algebra-homomorphisms.

§21. Substitutions with restrictions

Let R be a ring. Let Y be given where

$$\begin{cases} \text{either } Y \text{ is an indeterminate string over } R \\ \text{or } Y \text{ is an indeterminate net over } R \end{cases}$$

and let t and t' be given where

$$\begin{cases} t \text{ and } t' \text{ are string-restrictions in case } Y \text{ is a string} \\ \text{whereas } t \text{ and } t' \text{ are net-restrictions in case } Y \text{ is a net.} \end{cases}$$

Given any $\bar{Y}$ where

$$\begin{cases} \bar{Y} \in R[Y](o(Y)) \text{ in case } Y \text{ is a string} \\ \text{whereas } \bar{Y} \in R[Y](\ell(Y)) \text{ in case } Y \text{ is a net} \end{cases}$$

we define

$$\mathrm{Sub}[R, Y\langle t\rangle = \bar{Y}] : R[Y\langle t\rangle] \to R[Y]$$

to be the R-algebra-homomorphism induced by $\mathrm{Sub}[R, Y = \bar{Y}]$ and for any $R_0 \subset R$ with $0 \in R_0$ we put

$$R_0[\bar{Y}\langle t\rangle] = \mathrm{Sub}[R, Y = \bar{Y}](R_0[Y\langle t\rangle])$$

and we define

$$\mathrm{Sub}[R, Y\langle t\rangle = \bar{Y}]^* : R[Y\langle t\rangle] \to R[\bar{Y}\langle t\rangle]$$

to be the R-algebra-epimorphism induced by $\mathrm{Sub}[R, Y = \bar{Y}]$, and we define

$$\mathrm{Sub}[R, Y = \bar{Y}\langle t'\rangle]: R[Y] \to R[Y]$$

to be the unique R-algebra-homomorphism such that

$$\mathrm{Sub}[R, Y = \bar{Y}\langle t'\rangle](F) = \begin{cases} F & \text{for all } F \in R[Y\langle \neq t'\rangle] \\ \mathrm{Sub}[R, Y = \bar{Y}](F) & \text{for all } F \in R[Y\langle t'\rangle] \end{cases}$$

and we define

$$\mathrm{Sub}[R, Y\langle t\rangle = \bar{Y}\langle t'\rangle]: R[Y\langle t\rangle] \to R[Y]$$

to be the R-algebra-homomorphism induced by $\mathrm{Sub}[R, Y = \bar{Y}\langle t'\rangle]$, and we define

$$\mathrm{Sub}[R, Y = \bar{Y}\langle t'\rangle]^*: R[Y] \to \mathrm{Sub}[R, Y = \bar{Y}\langle t'\rangle](R[Y])$$

and

$$\mathrm{Sub}[R, Y\langle t\rangle = \bar{Y}\langle t'\rangle]^*: R[Y\langle t\rangle] \to \mathrm{Sub}[R, Y = \bar{Y}\langle t'\rangle](R[Y\langle t\rangle])$$

to be the R-algebra-epimorphisms induced by $\mathrm{Sub}[R, Y = \bar{Y}\langle t'\rangle]$.

Given any y where

$$\begin{cases} y \in R(o(Y)) & \text{in case } Y \text{ is a string} \\ \text{whereas } y \in R(\ell(Y)) & \text{in case } Y \text{ is a net} \end{cases}$$

we define

$$\mathrm{sub}[R, Y\langle t\rangle = y]: R[Y\langle t\rangle] \to R$$

to be the R-algebra-epimorphism induced by $\mathrm{Sub}[R, Y = y]$ and we note that then

$$\mathrm{sub}[R,\ Y\langle t\rangle = y] = \mathrm{Sub}[R,\ Y\langle t\rangle = y]^*.$$

Given any pseudomorphism $g\colon R \to R'$ and given any Y' where

$$\begin{cases} Y' \in R'[Y](o(Y)) & \text{in case } Y \text{ is a string} \\ \text{whereas } Y' \in R'[Y](\ell(Y)) & \text{in case } Y \text{ is a net} \end{cases}$$

we define

$$\mathrm{Sub}[g,\ Y\langle t\rangle = Y']\colon R[Y\langle t\rangle] \to R'[Y]$$

to be the pseudomorphism induced by $\mathrm{Sub}[g, Y = Y']$ and we define

$$\mathrm{Sub}[g,\ Y\langle t\rangle = Y']^*\colon R[Y\langle t\rangle] \to g(R)[Y'\langle t\rangle]$$

to be the surjective map induced by $\mathrm{Sub}[g, Y = Y']$, and we define

$$\mathrm{Sub}[g, Y = Y'\langle t'\rangle]\colon R[Y] \to R'[Y]$$

to be the pseudomorphism obtained by putting, for all $F \in R[Y]$, $\mathrm{Sub}[g,\ Y = Y'\langle t'\rangle](F)$

$$= \begin{cases} \displaystyle\sum_{j \in Z(o(Y),t')} \mathrm{Sub}[g,\ Y](F\langle \mathrm{supt}(o(Y),t')\rangle[j])Y'^{j} & \text{if } Y \text{ is a string} \\ \displaystyle\sum_{j \in Z(\ell(Y),t')} \mathrm{Sub}[g,Y](F\langle \mathrm{supt}(\ell(Y),t')\rangle[j])Y'^{j} & \text{if } Y \text{ is a net} \end{cases}$$

and we define

$$\mathrm{Sub}[g,\ Y\langle t\rangle = Y'\langle t'\rangle]\colon R[Y\langle t\rangle] \to R'[Y]$$

to be the pseudomorphism induced by $\mathrm{Sub}[g, Y = Y'\langle t'\rangle]$ and we define

$$\mathrm{Sub}[g, Y = Y'\langle t'\rangle]^*: R[Y] \to \mathrm{Sub}[g, Y = Y'\langle t'\rangle](R[Y])$$

and

$$\mathrm{Sub}[g, Y\langle t\rangle = Y'\langle t'\rangle]^*: R[Y\langle t\rangle] \to \mathrm{Sub}[g, Y = Y'\langle t'\rangle](R[Y\langle t\rangle])$$

to be the surjective maps induced by $\mathrm{Sub}[g, Y = Y'\langle t'\rangle]$ and we observe that: if g is a ring-homomorphism then the above defined six maps are ring-homomorphisms and out of them the three unstarred ones are g-algebra-homomorphisms.

Given any pseudomorphism $g: R \to R'$ and given any y' where

$$\begin{cases} y' \in R'(o(Y)) & \text{in case } Y \text{ is a string} \\ \text{whereas } y' \in R'(\ell(Y)) & \text{in case } Y \text{ is a net} \end{cases}$$

we define

$$\mathrm{sub}[g, Y\langle t\rangle = y']: R[Y\langle t\rangle] \to R'$$

to be the pseudomorphism induced by $\mathrm{Sub}[g, Y = y']$ and we define

$$\mathrm{sub}[g, Y\langle t\rangle = y']^*: R[Y\langle t\rangle] \to g(R)[y'\langle t\rangle]$$

to be the surjective map induced by $\mathrm{Sub}[g, Y = y']$ and we note that then

$$\mathrm{sub}[g, Y\langle t\rangle = y']^* = \mathrm{Sub}[g, Y\langle t\rangle = y']^*$$

and we observe that: if g is a ring-homomorphism then $\text{sub}[g, Y\langle t\rangle = y']$ and $\text{sub}[g, Y\langle t\rangle = y']^*$ are ring-homomorphisms and $\text{sub}[g, Y\langle t\rangle = y']$ is a g-algebra-homomorphism.

Given any pseudomorphism $g: R \to R'$ we define

$$\text{Sub}[g, Y\langle t\rangle]_Q : R[Y\langle t\rangle]_Q \to R'[Y]_Q$$

and

$$\text{Sub}[g, Y\langle t\rangle] : R[Y\langle t\rangle] \to R'[Y]$$

to be the pseudomorphisms induced by $\text{Sub}[g,Y]_Q$ and we define

$$\text{Sub}[g, Y\langle t\rangle]_Q^* : R[Y\langle t\rangle]_Q \to g(R)[Y\langle t\rangle]_Q$$

and

$$\text{Sub}[g, Y\langle t\rangle]^* : R[Y\langle t\rangle] \to g(R)[Y\langle t\rangle]$$

to be the surjective maps induced by $\text{Sub}[g,Y]_Q$ and we note that then

$$\text{Sub}[g, Y\langle t\rangle] = \text{Sub}[g, Y\langle t\rangle = Y] \quad \text{and} \quad \text{Sub}[g, Y\langle t\rangle]^* = \text{Sub}[g, Y\langle t\rangle = Y]^*$$

and we observe that: if g is a ring-homomorphism then $\text{Sub}[g, Y\langle t\rangle]_Q$ and $\text{Sub}[g, Y\langle t\rangle]_Q^*$ are ring-homomorphisms and $\text{Sub}[g, Y\langle t\rangle]_Q$ is a g-algebra-homomorphism.

We define

$$\text{Sub}[R, Y\langle t\rangle = 0]_Q : R[Y\langle t\rangle]_Q \to R[Y]_Q$$

and

$$\mathrm{Sub}[R, Y\langle t\rangle = 0]: R[Y\langle t\rangle] \to R[Y]$$

to be the R-algebra-homomorphisms induced by $\mathrm{Sub}[R,Y=0]_Q$ and we define

$$\mathrm{Sub}[R, Y\langle t\rangle = 0]^*_Q: R[Y\langle t\rangle]_Q \to R$$

$$\mathrm{Sub}[R, Y\langle t\rangle = 0]^*: R[Y\langle t\rangle] \to R$$

and

$$\mathrm{sub}[R, Y\langle t\rangle = 0]: R[Y\langle t\rangle] \to R$$

to be the R-algebra-epimorphisms induced by $\mathrm{Sub}[R,Y=0]_Q$ and we note that then

$$\mathrm{sub}[R, Y\langle t\rangle = 0] = \mathrm{Sub}[R, Y\langle t\rangle = 0]^*.$$

We define

$$\mathrm{Sub}[R,Y=0\langle t'\rangle]_Q: R[Y]_Q \to R[Y]_Q$$

to be the R-algebra-homomorphism obtained by putting, for all $F \in R[Y]_Q$,

$$\mathrm{Sub}[R, Y=0\langle t\rangle]_Q(F) = \begin{cases} \sum_{j \in Q(o(Y),t')} F\langle \mathrm{supt}(o(Y),t')\rangle[j] & \text{if } Y \text{ is a string} \\ \sum_{j \in Q(\ell(Y),t')} F\langle \mathrm{supt}(\ell(Y),t')\rangle[j] & \text{if } Y \text{ is a net} \end{cases}$$

and we define

$$\mathrm{Sub}[R,Y=0\langle t'\rangle]: R[Y] \to R[Y]$$

to be the R-algebra-homomorphism induced by $\mathrm{Sub}[R, Y = 0\langle t'\rangle]_Q$ and we define

$$\mathrm{Sub}[R, Y = 0\langle t'\rangle]_Q^*: R[Y]_Q \to R[Y\langle \neq t'\rangle]_Q$$

and

$$\mathrm{Sub}[R, Y = 0\langle t'\rangle]^*: R[Y] \to R[Y\langle \neq t'\rangle]$$

to be the R-algebra-epimorphisms induced by $\mathrm{Sub}[R, Y = 0\langle t'\rangle]_Q$.

We define

$$\mathrm{Sub}[R, Y\langle t\rangle = 0\langle t'\rangle]_Q: R[Y\langle t\rangle]_Q \to R[Y]_Q$$

and

$$\mathrm{Sub}[R, Y\langle t\rangle = 0\langle t'\rangle]: R[Y\langle t\rangle] \to R[Y]$$

to be the R-algebra-homomorphisms induced by $\mathrm{Sub}[R, Y = 0\langle t'\rangle]_Q$ and we define

$$\mathrm{Sub}[R, Y\langle t\rangle = 0\langle t'\rangle]_Q^*: R[Y\langle t\rangle]_Q \to R[Y\langle t\rangle]_Q \cap R[Y\langle \neq t'\rangle]_Q$$

and

$$\mathrm{Sub}[R, Y\langle t\rangle = 0\langle t'\rangle]^*: R[Y\langle t\rangle] \to R[Y\langle t\rangle] \cap R[Y\langle \neq t'\rangle]$$

to be the R-algebra-epimorphisms induced by $\mathrm{Sub}[R, Y = 0\langle t'\rangle]_Q$.

Given any pseudomorphism $g: R \to R'$ we define

$$\mathrm{Sub}[g, Y\langle t\rangle = 0]_Q: R[Y\langle t\rangle]_Q \to R'[Y]_Q$$

$$\mathrm{Sub}[g, Y\langle t\rangle = 0]: R[Y\langle t\rangle] \to R'[Y]$$

and

$$\mathrm{sub}[g,Y\langle t\rangle = 0]: R[Y\langle t\rangle] \to R'$$

to be the pseudomorphisms induced by $\mathrm{Sub}[g,Y=0]_Q$ and we define

$$\mathrm{Sub}[g,Y\langle t\rangle = 0]_Q^*: R[Y\langle t\rangle]_Q \to g(R)$$

$$\mathrm{Sub}[g,Y\langle t\rangle = 0]^*: R[Y\langle t\rangle] \to g(R)$$

and

$$\mathrm{sub}[g,Y\langle t\rangle = 0]^*: R[Y\langle t\rangle] \to g(R)$$

to be the surjective maps induced by $\mathrm{Sub}[g,Y=0]_Q$ and we note that then

$$\mathrm{sub}[g,Y\langle t\rangle = 0]^* = \mathrm{Sub}[g,Y\langle t\rangle = 0]^*$$

and we observe that: if g is a ring-homomorphism the above defined six maps are ring-homomorphisms and out of them the three unstarred one are g-algebra-homomorphisms.

Given any pseudomorphism $g: R \to R'$ we define the pseudomorphism

$$\mathrm{Sub}[g,Y = 0\langle t'\rangle]_Q: R[Y]_Q \to R'[Y]_Q$$

to be the composition

$$R[Y]_Q \xrightarrow{\mathrm{Sub}[g,Y]_Q} R'[Y]_Q \xrightarrow{\mathrm{Sub}[R',Y=0\langle t'\rangle]_Q} R'[Y]_Q$$

and we define

$$\mathrm{Sub}[g, Y = 0\langle t'\rangle] : R[Y] \to R'[Y]$$

$$\mathrm{Sub}[g, Y\langle t\rangle = 0\langle t'\rangle]_Q : R[Y\langle t\rangle]_Q \to R'[Y]_Q$$

and

$$\mathrm{Sub}[g, Y\langle t\rangle = 0\langle t'\rangle] : R[Y\langle t\rangle] \to R'[Y]$$

to be the pseudomorphisms induced by $\mathrm{Sub}[g, Y = 0\langle t'\rangle]_Q$ and we define

$$\mathrm{Sub}[g, Y = 0\langle t'\rangle]_Q^* : R[Y]_Q \to g(R)[Y\langle \neq t'\rangle]_Q$$

$$\mathrm{Sub}[g, Y = 0\langle t'\rangle]^* : R[Y] \to g(R)[Y\langle \neq t'\rangle]$$

$$\mathrm{Sub}[g, Y\langle t\rangle = 0\langle t'\rangle]_Q^* : R[Y\langle t\rangle]_Q \to g(R)[Y\langle t\rangle]_Q \cap g(R)[Y\langle \neq t'\rangle]_Q$$

and

$$\mathrm{Sub}[g, Y\langle t\rangle = 0\langle t'\rangle]^* : R[Y\langle t\rangle] \to g(R)[Y\langle t\rangle] \cap g(R)[Y\langle \neq t'\rangle]$$

to be the surjective maps induced by $\mathrm{Sub}[g, Y = 0\langle t'\rangle]_Q$ and we observe that: if g is a ring-homomorphism then the above defined eight maps are ring-homomorphisms and out of them the four unstarred ones are g-algebra-homomorphisms.

§22. Coordinate nets and Monic polynomials

Let R be a ring. Let Y be an indeterminate net over R. Let e be a prechip with $\ell(e) = \ell(Y)$.

Given any $B' \in [1, o(\ell(e))]$, by an $(R, Y\langle B'\rangle, e)$-coordinate-net we mean an $R[Y]$-net $\bar{Y}$ with $\ell(\bar{Y}) = \ell(Y)$ such that:

$$\bar{Y}(B,C) = Y(B,C) \quad \text{for} \quad \begin{cases} 1 \le B \le B'-1 \\ 1 \le C \le B(\ell(e)) \end{cases}$$

and

$$\bar{Y}(B,C) = Y(B,C) \quad \text{for} \quad \begin{cases} B' \le B \le o(\ell(e)) \\ 1 \le C \le T(-1,B,\ell(e)) \end{cases}$$

and

$$\bar{Y}(B,C) \in Y\langle B',B,T[\ell(e)]\rangle^{1}_{(R,e=)} \quad \text{for} \quad \begin{cases} B' \le B \le o(\ell(e))-1 \\ T(-1,B,\ell(e)) + 1 \le C \le B(\ell(e)) \end{cases}$$

and

$$\left.\begin{array}{l} \mathrm{Sub}[R, Y = 0\langle B+1\rangle](\bar{Y}(B)) \\ \text{is a free } R\text{-basis of } Y\langle 1,B\rangle^{1}_{(R=)} \end{array}\right\} \text{ for } B' \le B \le o(\ell(e))-1$$

Given any $B' \in [1, o(\ell(e))]$ we put

$$\mathrm{Coord}(R, Y\langle B'\rangle, e) = \text{the set of all } (R, Y\langle B'\rangle, e)\text{-coordinate-nets.}$$

Given any $B' \in [1,o(\ell(e))-1]$ and given any $B \in [B',o(\ell(e))-1]$ and $u \in Q$, we define

$$\mathrm{Nonmon}(R,Y\langle B',B\rangle,e=u)$$

$$= \bigcup_{\bar{Y}\in \mathrm{Coord}(R,Y\langle B'\rangle,e)} \mathrm{Sub}[R,Y=\bar{Y}]^{-1}(Y\langle B',B,T[\ell(e),B]\rangle^{u}_{(R,e=)})$$

and

$$\mathrm{Nonmon}((R,Y\langle B',B\rangle,e=u))$$

$$= \bigcup_{\bar{Y}\in \mathrm{Coord}(R,Y\langle B'\rangle,e)} \mathrm{Sub}[R,Y=\bar{Y}]^{-1}(Y\langle B',B,T[\ell(e),B]\rangle^{u}_{((R,e=))})$$

and we define

$$\mathrm{Mon}(R,Y\langle B',B\rangle,e=u)$$

$$= \mathrm{Iso}(R,Y\langle B',B,T[\ell(e)]\rangle,e=u)\backslash \mathrm{Nonmon}(R,Y\langle B',B\rangle,e=u)$$

and

$$\mathrm{Mon}((R,Y\langle B',B\rangle,e=u))$$

$$= \mathrm{Iso}((R,Y\langle B',B,T[\ell(e)]\rangle,e=u))\backslash \mathrm{Nonmon}((R,Y\langle B',B\rangle,e=u)).$$

By an (R,Y,e)-coordinate-net we mean an $(R,Y\langle 1\rangle,e)$-coordinate-net. Finally we put

$\mathrm{Coord}(R,Y,e) = \mathrm{Coord}(R,Y\langle 1\rangle,e)$.

Here we have used, and we shall continue to use, the following obvious conventions for sets S_1 and S_1' and a (set-theoretic) map $g: S \to S'$ where $S \subset S_1$ and $S' \subset S_1'$. If x is any S-string then by $g(x)$ we denote the $g(S)$-string such that $o(g(x)) = o(x)$ and $g(x)(c) = g(x(c))$ for all $c \in [1,o(x)]$; we may also regard $g(x)$ to be an S'-string as well as an S_1'-string. If S_0 is any set of S_1-strings then by $g(S_0)$ we denote the set $\{g(x): x \in S_0$ with $x(c) \in S$ for all $c \in [1,o(x)]\}$ of $g(S)$-strings; we may also regard $g(S_0)$ to be a set of S'-strings as well as a set of S_1'-strings. If S_0' is any set of S_1'-strings then by $g^{-1}(S_0')$ we denote the set of all S-strings x such that $g(x) \in S_0'$; we may also regard $g^{-1}(S_0')$ to be a set of S_1-strings. Similarly for S-nets, sets of S_1-nets, sets of S_1'-nets, S-webs, sets of S_1-webs, and sets of S_1'-webs, where the concept of webs is to be defined later.

We have also used, and we shall continue to use, the following obvious conventions for an S-module N and S-submodule N_0 of N where S is a ring. If x is an N-string then we say that x is a free S-basis of N_0 to mean that the indexed family $x(c)_{1 \le c \le o(x)}$ is a free S-basis of N_0. Similarly for N-nets and N-webs, where the concept of webs is to be defined later.

§23. Graded ring of a ring at an ideal

Let R be a ring and let I be an ideal in R.

By $gr(R,I)$ we denote the graded ring of R at I, i.e.,

$$gr(R,I) = \text{the external direct sum } \sum_{n=0}^{\infty} I^n/I^{n+1} \text{ with } I^0 = R,$$

and for every $n \in Z$ we define the map

$$gr^n[R,I]: I^n \to gr(R,I)$$

to be the composition of the natural maps

$$I^n \to I^n/I^{n+1} \to gr(R,I)$$

and we put

$$gr^n(R,I) = gr^n[R,I](I^n) .$$

We observe that now

$$gr(R,I) = \text{the (internal) direct sum } \sum_{n=0}^{\infty} gr^n(R,I).$$

We put

$$res(R,I) = gr^o(R,I)$$

and we define

$$res[R,I]: R \to res(R,I)$$

to be the ring-epimorphism induced by $gr^o[R,I]$. For any $f \in R$ we define

$$\mathrm{ord}[R,I](f) = \max\{n \in Z: f \in I^n\} \quad \text{where} \quad I_0 = R$$

and we note that

$$\mathrm{ord}[R,I](f) = \infty \Leftrightarrow f \in \bigcap_{n=1}^{\infty} I^n.$$

For any $f' \subset R$ we define

$$\mathrm{ord}[R,I](f') = \{\mathrm{ord}[R,I](f): f \in f'\}$$

and

$$\mathrm{ord}[R,I]((f')) = \min \mathrm{ord}[R,I](f')$$

and we note that:

$$\mathrm{ord}[R,I]((f')) = \infty \Leftrightarrow f' \subset \bigcap_{n=1}^{\infty} I^n.$$

We define the map

$$\mathrm{gr}[R,I]: R \to \mathrm{gr}(R,I)$$

by putting, for any $f \in R$,

$$\mathrm{gr}[R,I](f) = \begin{cases} \mathrm{gr}^n[R,I](f) & \text{if} \quad \mathrm{ord}[R,I](f) = n \neq \infty \\ 0 & \text{if} \quad \mathrm{ord}[R,I](f) = \infty. \end{cases}$$

We note that for any $f' \subset R$ we now have

$$\mathrm{gr}[R,I](f') = \{\mathrm{gr}[R,I](f): f \in f'\}$$

and we define

$$\mathrm{gr}[R,I]((f')) = \text{the ideal in } \mathrm{gr}(R,I) \text{ generated by } \mathrm{gr}[R,I](f')$$

and we observe that $gr[R,I]((f'))$ is then a homogeneous ideal in $gr(R)$.

Given any ideal J in R with $J \subset I$ we define

$$res[(R,I),J]: res(R,J) \to res(R,I)$$

to be the unique ring-epimorphism which makes the triangle

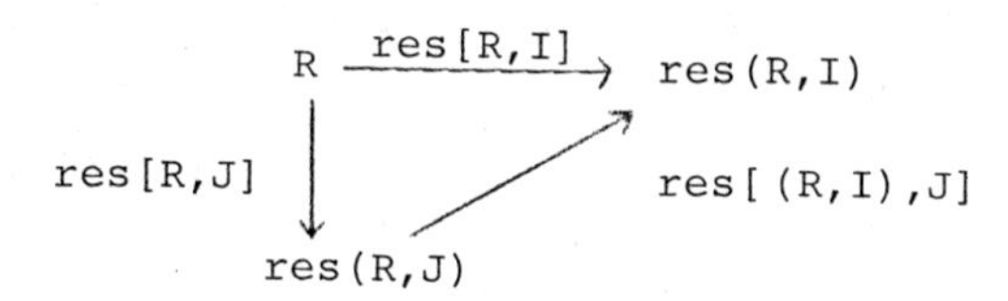

commutative.

§24. Graded ring of a ring

Let R be a ring. We define

$$M(R) = \text{the intersection of all maximal ideals in } R.$$

We put

$$gr(R) = gr(R,M(R)) \quad \text{and} \quad gr[R] = gr[R,M(R)]$$

and for every $n \in Z$ we put

$$gr^n[R] = gr^n[R,M(R)] \quad \text{and} \quad gr^n(R) = gr^n(R,M(R))$$

and we put

$$res[R] = res[R,M(R)] \quad \text{and} \quad res(R) = res(R,M(R))$$

and for any $f \in R$ we put

$$ord[R](f) = ord[R,M(R)](f)$$

and for any $f' \subset R$ we put

$$ord[R](f') = ord[R,M(R)](f')$$

and

$$ord[R]((f')) = ord[R,M(R)]((f'))$$

and

$$gr[R]((f')) = gr[R,M(R)]((f'))$$

and for any ideal J in R with $J \subset M(R)$ we put

$$res[(R),J] = res[(R,M(R)),J].$$

§25. Graded rings at strings and nets and the notions of separatedness and regularity for strings and nets

Let R be a ring. Let y be given where

$$\begin{cases} \text{either } y \text{ is an } R\text{-string} \\ \text{or } y \text{ is an } R\text{-net} \end{cases}$$

and let Y be given where

$$\begin{cases} Y \text{ is an indeterminate string over } R \text{ with } o(Y) = o(y) \\ \qquad\qquad \text{in case } y \text{ is an } R\text{-string} \\ \text{whereas } Y \text{ is an indeterminate net over } R \text{ with } \ell(Y) = \ell(y) \\ \qquad\qquad \text{in case } y \text{ is an } R\text{-net}. \end{cases}$$

We put

$$gr(R,y) = gr(R,y_R^1) \quad \text{and} \quad gr[R,y] = gr[R,y_R^1]$$

and for every $n \in Z$ we put

$$gr^n[R,y] = gr^n[R,y_R^1] \quad \text{and} \quad gr^n(R,y) = gr^n(R,y_R^1)$$

and we put

$$res[R,y] = res[R,y_R^1] \quad \text{and} \quad res(R,y) = res(R,y_R^1)$$

and for any $f \in R$ we put

$$ord[R,y](f) = ord[R,y_R^1](f)$$

and for any $f' \subset R$ we put

$$\mathrm{ord}[R,y](f') = \mathrm{ord}[R,y_R^1](f')$$

and

$$\mathrm{ord}[R,y]((f')) = \mathrm{ord}[R,y_R^1]((f'))$$

and

$$\mathrm{gr}[R,y]((f')) = \mathrm{gr}[R,y_R^1]((f')).$$

If $y_R^1 \subset M(R)$ then we put

$$\mathrm{res}[(R),y] = \mathrm{res}[(R),y_R^1].$$

If I is an ideal in R with $y_R^1 \subset I$ then we put

$$\mathrm{res}[(R,I),y] = \mathrm{res}[(R,I),y_R^1].$$

If $\bar{y}$ is an R-string with $y_R^1 \subset \bar{y}_R^1$ or $\bar{y}$ is an R-net with $y_R^1 \subset \bar{y}_R^1$ then we put

$$\mathrm{res}[(R,\bar{y}),y] = \mathrm{res}[(R,\bar{y}_R^1),y_R^1].$$

If $\bar{y}$ is an R-string and $\bar{t}$ is a string-restriction with $y_R^1 \subset \bar{y}\langle\bar{t}\rangle_R^1$ or $\bar{y}$ is an R-net and $\bar{t}$ is a net-restriction with $y_R^1 \subset \bar{y}\langle\bar{t}\rangle_R^1$ then we put

$$\mathrm{res}[(R,\bar{y}\langle\bar{t}\rangle),y] = \mathrm{res}[(R,\bar{y}\langle\bar{t}\rangle_R^1),y_R^1].$$

If J is an ideal in R with $J \subset y_R^1$ then we put

$$\mathrm{res}[(R,y),J] = \mathrm{res}[(R,y_R^1),J].$$

We define

$$gr[R,Y=y]: res(R,y)[Y] \to gr(R,y)$$

to be the unique $res(R,y)$-algebra-epimorphism such that

$$\begin{cases} gr[R,Y=y](Y(c)) = gr^1[R,y](y(c)) \text{ for all } c \in [1,o(y)] \\ \qquad \text{in case } y \text{ is an } R\text{-string} \end{cases}$$

whereas

$$\begin{cases} gr[R,Y=y](Y(b,c)) = gr^1[R,y](y(b,c)) \text{ for all } (b,c) \in supt(\ell(y)) \\ \qquad \text{in case } y \text{ is an } R\text{-net.} \end{cases}$$

Now let t be given where

$$\begin{cases} t \text{ is a string-restriction in case } y \text{ is an } R\text{-string} \\ \text{whereas } t \text{ is a net-restriction in case } y \text{ is an } R\text{-net.} \end{cases}$$

We put

$$gr(R,y\langle t\rangle) = gr(R,y\langle t\rangle^1_R) \quad \text{and} \quad gr[R,y\langle t\rangle] = gr[R,y\langle t\rangle^1_R]$$

and for every $n \in Z$ we put

$$gr^n[R,y\langle t\rangle] = gr^n[R,y\langle t\rangle^1_R] \quad \text{and} \quad gr^n(R,y\langle t\rangle) = gr^n(R,y\langle t\rangle^1_R)$$

and we put

$$res[R,y\langle t\rangle] = res[R,y\langle t\rangle^1_R] \quad \text{and} \quad res(R,y\langle t\rangle) = res(R,y\langle t\rangle^1_R)$$

and for any $f \in R$ we put

$$ord[R,y\langle t\rangle](f) = ord[R,y\langle t\rangle^1_R](f) \quad .$$

and for any $f' \subset R$ we put

$$\mathrm{ord}[R,y\langle t\rangle](f') = \mathrm{ord}[R,y\langle t\rangle_R^1](f')$$

and

$$\mathrm{ord}[R,y\langle t\rangle]((f')) = \mathrm{ord}[R,y\langle t\rangle_R^1]((f'))$$

and

$$\mathrm{gr}[R,y\langle t\rangle]((f')) = \mathrm{gr}[R,y\langle t\rangle_R^1]((f')).$$

If $y\langle t\rangle_R^1 \subset M(R)$ then we put

$$\mathrm{res}[(R),y\langle t\rangle] = \mathrm{res}[(R),y\langle t\rangle_R^1].$$

If I is an ideal in R with $y\langle t\rangle_R^1 \subset I$ then we put

$$\mathrm{res}[(R,I),y\langle t\rangle] = \mathrm{res}[(R,I),y\langle t\rangle_R^1].$$

If $\bar{y}$ is an R-string with $y\langle t\rangle_R^1 \subset \bar{y}_R^1$ or $\bar{y}$ is an R-net with $y\langle t\rangle_R^1 \subset \bar{y}_R^1$ then we put

$$\mathrm{res}[(R,\bar{y}),y\langle t\rangle] = \mathrm{res}[(R,\bar{y}_R^1),y\langle t\rangle_R^1].$$

If $\bar{y}$ is an R-string and $\bar{t}$ is a string-restriction with $y\langle t\rangle_R^1 \subset \bar{y}\langle\bar{t}\rangle_R^1$ or $\bar{y}$ is an R-net and $\bar{t}$ is a net-restriction with $y\langle t\rangle_R^1 \subset \bar{y}\langle\bar{t}\rangle_R^1$ then we put

$$\mathrm{res}[(R,\bar{y}\langle\bar{t}\rangle),y\langle t\rangle] = \mathrm{res}[(R,\bar{y}\langle\bar{t}\rangle_R^1),y\langle t\rangle_R^1].$$

If J is an ideal in R with $J \subset y\langle t\rangle_R^1$ then we put

$$\mathrm{res}[(R,y\langle t\rangle),J] = \mathrm{res}[(R,y\langle t\rangle_R^1),J].$$

We define

$$gr[R,\ Y\langle t\rangle = y]:\ res(R,y\langle t\rangle)[Y\langle t\rangle] \to gr(R,y\langle t\rangle)$$

to be the unique $res(R,y\langle t\rangle)$-algebra-epimorphism such that

$$\begin{cases} gr[R,\ Y\langle t\rangle = y](Y(c)) = gr^1[R,y\langle t\rangle](y(c)) \text{ for all } c \in supt(o(y),t) \\ \qquad\qquad \text{in case } y \text{ is an R-string} \end{cases}$$

whereas

$$\begin{cases} gr[R,\ Y\langle t\rangle = y](Y(b,c)) = gr^1[R,y\langle t\rangle](y(b,c)) \text{ for all} \\ \qquad (b,c) \in supt(\ell(y),t) \text{ in case } y \text{ is an R-net.} \end{cases}$$

DEFINITION 1. In case y is an R-string, we say that y is R-<u>separated</u> to mean that

$$y_R^1 \neq R$$

and

$$\bigcap_{n=1}^{\infty} (y\langle[1,c-1]\rangle_R^1 + y_R^n) = y\langle[1,c-1]\rangle_R^1 \quad \text{for all } c \in [1,o(y)].$$

DEFINITION 2. In case y is an R-string, we say that y is R-<u>regular</u> to mean that

$$y_R^1 \neq R$$

and

$$\begin{cases} \text{for every } c \in [1,o(y)] \text{ and every } z \in R \text{ with } z \notin y\langle[1,c-1]\rangle_R^1 \\ \text{we have } zy(c) \notin y\langle[1,c-1]\rangle_R^1 . \end{cases}$$

DEFINITION 3. In case y is an R-string, we say that y is R-ultraseparated (resp: R-ultraregular) to mean that for every bijection $H: [1,o(y)] \to [1,o(y)]$, upon letting $\tilde{H}(y)$ to be the R-string with $o(\tilde{H}(y)) = o(y)$ and $\tilde{H}(y)(c) = y(H(c))$ for all $c \in [1,o(y)]$, we have that $\tilde{H}(y)$ is R-separated (resp: R-regular).

DEFINITION 4. In case y is an R-string, we say that y is R-superregular to mean that y is R-ultraseparated and y is R-ultraregular.

DEFINITION 5. In case y is an R-string, we say that $y\langle t\rangle$ is R-separated (resp: R-ultraseparated, R-regular, R-ultraregular, R-superregular) to mean that, upon letting $h: [1,\text{card}(\text{supt}(o(y),t))] \to \text{supt}(o(y),t)$ to be the unique order-preserving bijection and upon letting $\tilde{x}$ to be the R-string such that $o(\tilde{x}) = \text{card}(\text{supt}(o(y),t))$ and $\tilde{x}(c) = y(h(c))$ for all $c \in [1,\text{card}(\text{supt}(o(y),t))]$, we have that $\tilde{x}$ is R-separated (resp: R-ultraseparated, R-regular, R-ultraregular, R-superregular).

DEFINITION 6. In case y is an R-net, let $g: [1,\text{card}(\text{supt}(\ell(y),t))] \to \text{supt}(\ell(y),t)$ be the unique bijection such that for every c_1 and c_2 in $[1,\text{card}(\text{supt}(\ell(y),t))]$, upon letting $(B_i,C_i) = g(c_i)$ with $B_i \in [1,o(\ell(y))]$ and $C_i \in [1,B_i(\ell(y))]$ we have: $c_1 \le c_2 \Leftrightarrow$ either $B_1 = B_2$ and $C_1 \le C_2$ or $B_1 < B_2$; now let x be the R-string such that $o(x) = \text{card}(\text{supt}(\ell(y),t))$ and such that for every c in $[1,\text{card}(\text{supt}(\ell(y),t))]$, upon letting $(B,C) = g(c)$ with

$B \in [1,o(\ell(y))]$ and $C \in [1,B(\ell(y))]$, we have $x(c) = y(B,C)$; with this notation in mind, we say that $y\langle t\rangle$ is R-separated (resp: R-ultraseparated, R-regular, R-ultraregular, R-superregular) to mean that x is R-separated (resp: R-ultraseparated, R-regular, R-ultraregular, R-superregular).

DEFINITION 7. In case y is an R-net, we say that y is R-separated (resp: R-ultraseparated, R-regular, R-ultraregular, R-superregular) to mean that $y\langle 1\rangle$ is R-separated (resp: R-ultraseparated, R-regular, R-ultraregular, R-superregular).

DEFINITION 8. We say that $y\langle t\rangle$ is R-quasiregular to mean that $y\langle t\rangle^{1}_{R} \nmid R$ and $gr[R, Y\langle t\rangle = y]$ is injective. We say that $y\langle t\rangle$ is R-ultraquasiregular to mean that $y\langle t'\rangle$ is R-quasiregular for every t'

$$\begin{cases} \text{where } t' \subset \text{supt}(o(y),t) \text{ in case } y \text{ is an R-string} \\ \text{whereas } t' \subset \text{supt}(\ell(y),t) \text{ in case } y \text{ is an R-net.} \end{cases}$$

We say that y is R-quasiregular (resp: R-ultraquasiregular) to mean that $y\langle 1\rangle$ is R-quasiregular (resp: R-ultraquasiregular).

DEFINITION 9. Assume that $y\langle t\rangle$ is R-quasiregular. Then by

$$gr[R, y\langle t\rangle = Y]^{*}: R \to \text{res}(R,y\langle t\rangle)[Y\langle t\rangle]$$

we denote the unique surjective map which makes the triangle

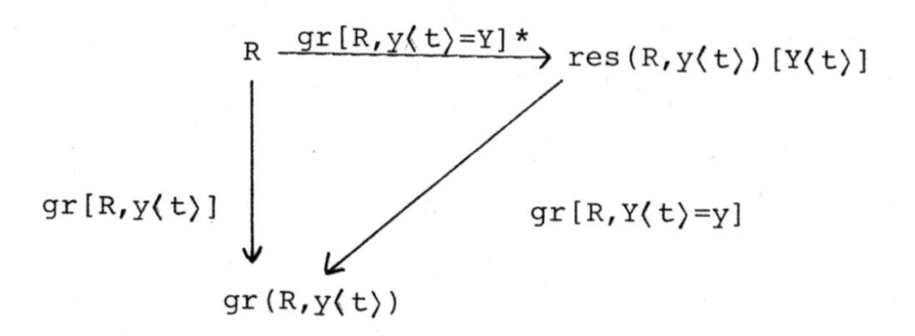

commutative. Also we define

$$gr[R,\ y\langle t\rangle=Y]:\ R \to res(R,y\langle t\rangle)[Y]$$

to be the map induced by $gr[R,\ y\langle t\rangle=Y]^*$. For any $f' \subset R$ we define

$$gr[R,\ y\langle t\rangle = Y]^*((f')) = \text{the ideal in } res(R,\ y\langle t\rangle)[Y\langle t\rangle] \text{ generated by } gr[R,\ y\langle t\rangle=Y]^*(f')$$

and

$$gr[R,\ y\langle t\rangle=Y]((f')) = \text{the ideal in } res[R,\ y\langle t\rangle=Y](f') \text{ generated by } gr\ [R,\ y\langle t\rangle=Y](f').$$

For every $n \in Z$ we define

$$gr^n[R,\ y\langle t\rangle=Y]:\ y\langle t\rangle_R^n \to res(R,\ y\langle t\rangle)[Y]$$

to be the $res[R,\ y\langle t\rangle]$-homomorphism obtained by putting

$$gr^n[R,y\langle t\rangle=Y](f) = \begin{cases} gr[R,y\langle t\rangle=Y](f) & \text{for all } f \in y\langle t\rangle_R^n \backslash y\langle t\rangle_R^{n+1} \\ 0 & \text{for all } f \in y\langle t\rangle_R^{n+1} \end{cases}$$

and we define

$$gr^n[R,\ y\langle t\rangle=Y]^*:\ y\langle t\rangle_R^n \to Iso(res(R,\ y\langle t\rangle),Y\langle t\rangle,\ =n)$$

to be the $res[R, y\langle t\rangle]$-epimorphism induced by $gr^n[R, y\langle t\rangle = Y]$.
Given any I where

$$\begin{cases} \text{either } I \text{ is an ideal in } R \text{ with } y\langle t\rangle_R^1 \subset I \\ \text{or } I = \bar{x} \text{ where } \bar{x} \text{ is an R-string with } y\langle t\rangle_R^1 \subset \bar{x}_R^1 \\ \text{or } I = \bar{x}\langle\bar{t}\rangle \text{ where } \bar{x} \text{ is an R-string and} \\ \qquad\qquad \bar{t} \text{ is a string-restriction with } y\langle t\rangle_R^1 \subset \bar{x}\langle\bar{t}\rangle_R^1 \\ \text{or } I = \bar{y} \text{ where } \bar{y} \text{ is an R-net with } y\langle t\rangle_R^1 \subset \bar{y}_R^1 \\ \text{or } I = \bar{y}\langle\bar{t}\rangle \text{ where } \bar{y} \text{ is an R-net and} \\ \qquad\qquad \bar{t} \text{ is a net restriction with } y\langle t\rangle_R^1 \subset \bar{y}\langle\bar{t}\rangle_R^1 \end{cases}$$

we define

$$gr[(R,I), y\langle t\rangle = Y]: R \to res(R,I)[Y]$$

to be the composition of the maps

$$\begin{array}{ccc} R & \xrightarrow{gr[R, y\langle t\rangle = Y]} & res(R, y\langle t\rangle)[Y] \\ & & \Big\downarrow Sub[res[(R,I), y\langle t\rangle], Y] \\ & & res(R,I)[Y] \end{array}$$

and we define

$$gr[(R,I), y\langle t\rangle = Y]^*: R \to res(R,I)[Y\langle t\rangle]$$

to be the surjective map induced by $gr[(R,I), y\langle t\rangle = Y]$ and for any $f' \subset R$ we define

$gr[(R,I),y\langle t\rangle=Y]^*((f'))$ = the ideal in $res(R,I)[Y\langle t\rangle]$ generated by $gr[(R,I),y\langle t\rangle=Y]^*(f')$

and

$gr[(R,I),y\langle t\rangle=Y]((f'))$ = the ideal in $res(R,I)[Y]$ generated by $gr[(R,I),y\langle t\rangle=Y](f')$

and for any $n \in Z$ we define the $res[R,I]$-homomorphism

$$gr^n[(R,I),y\langle t\rangle=Y]: y\langle t\rangle_R^n \to res(R,I)[Y]$$

to be the composition of the maps

$$Y\langle t\rangle_R^n \xrightarrow{gr^n[R,y\langle t\rangle=Y]} res(R,y\langle t\rangle)[Y] \xrightarrow{sub[res[(R,I),y\langle t\rangle],Y]} res(R,I)[Y]$$

and we define

$$gr^n[(R,I),y\langle t\rangle=Y]^*: y\langle t\rangle_R^n \to Iso(res(R,I),Y\langle t\rangle, =n)$$

to be the $res[R,I]$-epimorphism induced by $gr^n[(R,I),y\langle t\rangle=Y]$. If $y\langle t\rangle_R^1 \subset M(R)$ then we define the map

$$gr[(R),y\langle t\rangle=Y]: R \to res(R)[Y]$$

by putting

$$gr[(R),y\langle t\rangle=Y] = gr[(R,M(R)),y\langle t\rangle=Y]$$

and we define

$$gr[(R),y\langle t\rangle=Y]^*: R \to res(R)[Y\langle t\rangle]$$

to be the surjective map induced by $gr[(R),y\langle t\rangle=Y]$ and for any $f' \subset R$ we define

$$gr[(R),y\langle t\rangle=Y]^*((f')) = \text{the ideal in } res(R)[Y\langle t\rangle] \text{ generated by } gr[(R),y\langle t\rangle=Y]^*(f')$$

and

$$gr[(R),y\langle t\rangle=Y]((f')) = \text{the ideal in } res(R)[Y] \text{ generated by } gr[(R),y\langle t\rangle=Y](f')$$

and for any $n \in Z$ we define the $res[R]$-homomorphism

$$gr^n[(R),y\langle t\rangle=Y]: y\langle t\rangle_R^n \to res(R)[Y]$$

by putting

$$gr^n[(R),y\langle t\rangle=Y] = gr^n[(R,M(R)),y\langle t\rangle=Y]$$

and we define

$$gr^n[(R),y\langle t\rangle=Y]^*: y\langle t\rangle_R^n \to Iso(res(R),Y\langle t\rangle, =n)$$

to be the $res[R]$-epimorphism induced by $gr^n[(R),y\langle t\rangle=Y]$.

DEFINITION 10. If y is R-quasiregular then we verbatim take over the entire above material of Definition 9 after everywhere deleting $\langle t\rangle$.

LEMMA 1. If $y\langle t\rangle$ is R-quasiregular and if $f \in R$ and $g \in R$ are such that

$$ord[R,y\langle t\rangle]f \neq \infty$$

and

$$g = \begin{cases} y(c) \text{ for some } c \in \mathrm{supt}(o(y),t) \text{ in case } y \text{ is an R-string} \\ y(b,c) \text{ for some } (b,c) \in \mathrm{supt}(\ell(y),t) \text{ in case } y \text{ is an R-net} \end{cases}$$

then

$$\mathrm{ord}[R,y\langle t\rangle](fg) = 1 + \mathrm{ord}[R,y\langle t\rangle](f).$$

PROOF. Obvious.

LEMMA 2. Assume that $y\langle t\rangle$ is R-quasiregular. Let t' and t^* be disjoint sets such that

$$\begin{cases} t' \cup t^* = \mathrm{supt}(o(y),t) \text{ in case } y \text{ is an R-string} \\ \text{whereas } t' \cup t^* = \mathrm{supt}(\ell(y),t) \text{ in case } y \text{ is an R-net.} \end{cases}$$

Let $\overline{R} = \mathrm{res}(R,y\langle t^*\rangle)$ and $\overline{y} = \mathrm{res}[R,y\langle t^*\rangle](y)$. Then $\overline{y}\langle t'\rangle$ is $\overline{R}$-quasiregular.

PROOF. The case when y is an R-net follows from the case when y is an R-string. Also the case when $\mathrm{card}(t^*) = 0$ is obvious. Finally, when y is an R-string, the general case of $\mathrm{card}(t^*) \neq 0$ follows from the case of $\mathrm{card}(t^*) = 1$ by induction on $\mathrm{card}(t^*)$. So we may suppose that y is an R-string, $y\langle t\rangle$ is R-quasiregular, and $\mathrm{card}(t^*) = 1$. Let c be the element in $[1,o(y)]$ such that

$$t^* = \{c\}.$$

Let there be given any

$$\overline{H} \in \mathrm{Iso}(\overline{R},Y\langle t'\rangle, =v) \text{ with } 0 \neq v \in Z$$

such that

$$\text{sub}[\bar{R}, Y = \bar{y}](\bar{H}) = \bar{z} \quad \text{with} \quad \bar{z} \in \bar{y}\langle t' \rangle_R^{v+1}.$$

We shall show that then $\text{Sub}[\text{res}[\bar{R},\bar{y}\langle t' \rangle], Y](\bar{H}) = 0$ and this will complete the proof.

We can take

$$H \in \text{Iso}(R, Y\langle t' \rangle, =v) \quad \text{and} \quad z \in y\langle t \rangle_R^{v+1}$$

such that

$$\text{Sub}[\text{res}[R, y\langle t^* \rangle], Y](H) = \bar{H} \quad \text{and} \quad \text{res}[R, y\langle t^* \rangle](z) = \bar{z}.$$

Now

$$\text{sub}[R, Y = y](H) = z + fy(c) \quad \text{with} \quad f \in R$$

and in view of Lemma 1 we see that

$$\text{ord}[R, y\langle t \rangle](f) \geq v - 1$$

and hence

$$f = \text{sub}[R, Y = y](F) \quad \text{for some} \quad F \in \text{Iso}(R, Y\langle t \rangle, =v-1).$$

Let

$$\hat{F} = H - FY(c).$$

Then

$$\hat{F} \in \text{Iso}(R, Y\langle t \rangle, =v) \text{ and } \text{sub}[R, Y = y](\hat{F}) = z \in y\langle t \rangle_R^{v+1}$$

and hence by the R-quasiregularity of $y\langle t \rangle$ we must have

$$\text{Sub}[\text{res}[R, y\langle t \rangle], Y](\hat{F}) = 0.$$

Because $\hat{F} = H - FY\langle c\rangle$ and $H \in R[Y\langle t'\rangle]$ with $c \notin t'$, the above equation yields that

$$\mathrm{Sub}[\mathrm{res}[R,y\langle t\rangle],Y](H) = 0.$$

Therefore

$$\mathrm{Sub}[\mathrm{res}[\overline{R},\overline{y}\langle t'\rangle],Y](\overline{H}) = 0.$$

LEMMA 3. Assume that $y\langle t\rangle$ is R-separated and R-quasiregular. Then $y\langle t\rangle$ is R-regular.

PROOF. The case when y is an R-net follows from the case when y is an R-string. So we shall suppose that y is an R-string and we shall make induction on $\mathrm{card}(\mathrm{supt}(o(y),t))$. The assertion is obvious when $\mathrm{card}(\mathrm{supt}(o(y),t)) = 0$. So now let $\mathrm{card}(\mathrm{supt}(o(y),t)) \neq 0$ and assume that the assertion is true for all values of $\mathrm{card}(\mathrm{supt}(o(y),t))$ smaller than the given one. Let $c = \min \mathrm{supt}(o(y),t)$. Then, in view of Lemma 1, for every $0 \neq z \in R$ we have $zy(c) \neq 0$. Let $\overline{R} = \mathrm{res}(R,y\langle\{c\}\rangle)$ $\overline{y} = \mathrm{res}[R,y\langle\{c\}\rangle](y)$ and $t' = \mathrm{supt}(o(y),t)\setminus\{c\}$; then obviously $\overline{y}\langle t'\rangle$ is $\overline{R}$-separated, and by Lemma 2 we also see that $\overline{y}\langle t'\rangle$ is $\overline{R}$-quasiregular; since $\mathrm{card}(t') < \mathrm{card}(\mathrm{supt}(o(y),t))$, by the induction hypothesis it follows that $\overline{y}\langle t\rangle$ is $\overline{R}$-regular. Therefore $y\langle t\rangle$ is R-regular.

LEMMA 4. If $y\langle t\rangle$ is R-ultraseparated and R-quasiregular then $y\langle t\rangle$ is R-superregular.

PROOF. Follows from Lemma 3.

§26. Inner products and further notions of separatedness and regularity for strings

Let G' be a nonnegative ordered additive abelian semigroup. Let Q' be the rational completion of G', i.e., let Q' be the unique (upto G'-isomorphisms) divisible nonnegative ordered additive abelian oversemigroup of G' such that for every $u \in Q'$ we have $nu \in G'$ for some $0 \neq n \in Z$; (n depending on u). Let D be a G'-string.

For any $i \in Q(o(D))$ we define

$$\mathrm{inpo}(i,D) = \sum_{1 \le c \le o(D)} i(c)D(c)$$

and we note that

$$\mathrm{inpo}(i,D) \in Q'$$

and:

$$\text{if} \quad i \in Z(o(D)) \quad \text{then} \quad \mathrm{inpo}(i,D) \in G' .$$

For any $i' \subset Q(o(D))$ we put

$$\mathrm{inpo}(i',D) = \{\mathrm{inpo}(i,D) : i \in i'\} .$$

For any $G \subset Q$ we put

$$\mathrm{inpo}(G,D) = \mathrm{inpo}(G(o(D)),D)$$

and we note that:

$$\text{if} \quad G \subset Z \quad \text{then} \quad \mathrm{inpo}(G,D) \subset G' .$$

For any $G \subset Q$ and any string-restriction t we put

$$\text{inpo}(G,D\langle t\rangle) = \{\text{inpo}(i,D): i \in G(o(D)) \text{ with } \text{supt}(i) \subset \text{supt}(o(D),t)\}$$

and we note that:

$$\text{if } G \subset Z \text{ then } \text{inpo}(G,D\langle t\rangle) \subset G' .$$

For any $G \subset Q$ and any $u \in Q'$ and $P \in \{=,\geq,>,<,\leq\}$ we put

$$G(DPu) = \{i \in G(o(D)): \text{inpo}(i,D)Pu\} .$$

Recall that G' is <u>archimedian</u> means that for every $u \in G'$ and every $0 \neq u' \in G'$ we have $n(u,u')u' \geq u$ for some $n(u,u') \in Z$.

Now let there be given a string-restriction t.

Let

$$\hat{t} = \text{supt}(o(D),t) \cap \text{supt}(D)$$

and

$$t' = \{c' \in \hat{t}: \text{for every } c \in t \text{ we have } n(c,c')D(c') \geq D(c) \text{ for some } n(c,c') \in Z\}$$

and

$$t'' = \hat{t} \backslash t' .$$

We say that $D\langle t\rangle$ is <u>archimedian</u> to mean that $t'' = \emptyset$.
We note that

(1) $D\langle t\rangle$ is archimedian $\Leftrightarrow D\langle \hat{t}\rangle$ is archimedian

and we also note that

(2) if G' is archimedian then so are $D\langle t\rangle$ and $D\langle \hat{t}\rangle$.

We say that D is <u>archimedian</u> to mean that $D\langle 1\rangle$ is archimedian.

Now clearly there exists a unique sequence $t_1, t_2, \ldots, t_m$ of pairwise disjoint nonempty subsets of t'' with

$$m \in Z \quad \text{and} \quad t'' = \bigcup_{1\le r\le m} t_r$$

such that for $1 \le q \le m$ it is true that:

$$\begin{cases} \text{for every } c' \in t_q \text{ and every } c \in t_q \text{ we have} \\ n(c,c')D(c') \ge D(c) \quad \text{for some } n(c,c') \in Z \end{cases}$$

and

$$\begin{cases} \text{for every } c' \in t_q \text{ and every } c = \bigcup_{q+1\le r\le m} t_q \text{ we have} \\ D(c') > nD(c) \quad \text{for all } n \in Z. \end{cases}$$

Note that:

$$m = 0 \Leftrightarrow t'' = \emptyset .$$

Let R be a ring and let x be an R-string with $o(x) = o(D)$.

We say that $x\langle t\rangle$ is (R,D)-<u>preseparated</u> to mean that for $1 \le q \le m$ we have

$$\bigcap_{n=1}^{\infty} (x\langle t' \cup t_1 \cup \ldots \cup t_{q-1}\rangle)_R^1 + x\langle t_q\rangle^n) = x\langle t' \cup t_1 \cup \ldots \cup t_{q-1}\rangle_R^1$$

We note that

(3) $x\langle t\rangle$ is (R,D)-preseparated $\Leftrightarrow$ $x\langle \hat{t}\rangle$ is (R,D)-preseparated.

Let $h\colon [1,\mathrm{card}(\hat{t})] \to \hat{t}$ be the unique order-preserving bijection. Let $\tilde{x}$ be the R-string such that $o(\tilde{x}) = \mathrm{card}(\hat{t})$ and $\tilde{x}(c) = x(h(c))$ for all $c \in [1,\mathrm{card}(\hat{t})]$. Let $\tilde{D}$ be the G'-string such that $o(\tilde{D}) = o(\tilde{x})$ and $\tilde{D}(c) = D(h(c))$ for all $c \in [1,o(\tilde{x})]$.

We say that $x\langle t\rangle$ is (R,D)-<u>separated</u> to mean that $\tilde{x}\langle [1,c]\rangle$ is $(R,\tilde{D})$-preseparated for $1 \le c \le o(\tilde{x})$. We note that

(4) if $x\langle t\rangle$ is (R,D)-separated then $x\langle t\rangle$ is (R,D)-preseparated

and we note that

(5) $x\langle t\rangle$ is (R,D)-separated $\Leftrightarrow$ $x\langle \hat{t}\rangle$ is (R,D)-separated

and we also note that

(6) if either $D\langle t\rangle$ is archimedian or $D\langle \hat{t}\rangle$ is archimedian or $x\langle t\rangle$ is R-ultraseparated or $x\langle \hat{t}\rangle$ is R-ultraseparated then: $x\langle t\rangle$ is (R,D)-separated.

We say that $x\langle t\rangle$ is (R,D)-<u>regular</u> to mean that $x\langle t\rangle$ is R-regular and $x\langle t\rangle$ is (R,D)-separated. We note that

(7) $x\langle t\rangle$ is (R,D)-regular $\Leftrightarrow$ $x\langle \hat{t}\rangle$ is (R,D)-regular

and we also note that

$$(8)\quad \begin{cases} \text{if either } D\langle t\rangle \text{ is archimedian or } D\langle \hat{t}\rangle \text{ is archimedian} \\ \text{or } x\langle t\rangle \text{ is R-ultraseparated or } x\langle \hat{t}\rangle \text{ is R-ultraseparated} \\ \text{then: } x\langle t\rangle \text{ is (R,D)-regular} \Leftrightarrow x\langle \hat{t}\rangle \text{ is R-regular.} \end{cases}$$

We say that x is (R,D)-<u>regular</u> to mean that $x\langle 1\rangle$ is (R,D)-regular.

§27. Inner products and further notions of separatedness and regularity for nets.

Let G' be a nonnegative ordered additive abelian semigroup and let Q' be the rational completion of G'. Let E be a G'-net.

For any $j \in Q(\ell(E))$ we define

$$\operatorname{inpo}(j,E) = \sum_{\substack{1\le b\le o(\ell(E))\\ 1\le c\le b(\ell(E))}} j(b,c)E(b,c)$$

and we note that

$$\operatorname{inpo}(j,E) \in Q'$$

and

$$\text{if } j \in Z(\ell(E)) \text{ then } \operatorname{inpo}(j,E) \in G' .$$

For any $j' \subset Q(\ell(E))$ we put

$$\operatorname{inpo}(j',E) = \{\operatorname{inpo}(j,E) : j \in j'\} .$$

For any $G \subset Q$ we put

$$\operatorname{inpo}(G,E) = \operatorname{inpo}(G(\ell(E)),E)$$

and we note that:

$$\text{if } G \subset Z \text{ then } \operatorname{inpo}(G,E) \subset G' .$$

For any $G \subset Q$ and any net-restriction t we put

$\text{inpo}(G,E\langle t\rangle) = \{\text{inpo}(i,E): i \in G(\ell(E)) \text{ with } \text{supt}(i) \subset \text{supt}(\ell(E)),t)\}$

and we note that:

$$\text{if } G \subset Z \text{ then } \text{inpo}(G,E\langle t\rangle) \subset G' .$$

For any $G \subset Q$ and any $u \in Q'$ and $P \in \{=,\geq,>,<,\leq\}$ we put

$$G(EPu) = \{j \in G(\ell(E)): \text{inpo}(j,E)Pu\} .$$

Now let there be given a net-restriction t.

Let

$$\hat{t} = \text{supt}(\ell(E),t) \cap \text{supt}(E).$$

Let $g: [1,\text{card}(\hat{t})] \to \hat{t}$ be the unique bijection such that for every c_1 and c_2 in $[1,\text{card}(\hat{t})]$, upon letting $(B_i,C_i) = g(c_i)$ with $B_i \in [1,o(\ell(E))]$ and $C_i \in [1,B_i(\ell(E))]$, we have $c_1 \leq c_2 \Leftrightarrow$ either $B_1 = B_2$ and $C_1 \leq C_2$ or $B_1 < B_2$. Let D be the G'-string such that $o(D) = \text{card}(\hat{t})$ and such that for every c in $[1,\text{card}(\hat{t})]$, upon letting $(B,C) = g(c)$ with $B \in [1,o(\ell(E))]$ and $C \in [1,B(\ell(E))]$, we have $D(c) = E(B,C)$.

We say that $E\langle t\rangle$ is <u>archimedian</u> to mean that D is archimedian. We note that

(1) $\quad E\langle t\rangle$ is archimedian $\Leftrightarrow$ $E\langle\hat{t}\rangle$ is archemedian

and we also note that

(2) $\quad$ if G' is archimedian then so are $E\langle t\rangle$ and $E\langle\hat{t}\rangle$.

We say that E is <u>archimedian</u> to mean that $E\langle 1\rangle$ is archimedian.

Now let R be a ring and let y be an R-net with $\ell(y) = \ell(E)$.

Let x be the R-string such that $o(x) = o(D)$ and such that for every $c \in [1,o(D)]$, upon letting $(B,C) = g(c)$ with $B \in [1,o(\ell(y))]$ and $C \in [1,B(\ell(y))]$, we have $x(c) = y(B,C)$.

We say that $y\langle t\rangle$ is (R,E)-<u>preseparated</u> to mean that x is (R,D)-preseparated. We note that

(3) $y\langle t\rangle$ is (R,E)-preseparated $\Leftrightarrow$ $y\langle\hat{t}\rangle$ is (R,E)-preseparated.

We say that $y\langle t\rangle$ is (R,E)-<u>separated</u> to mean that x is (R,D)-separated. We note that

(4) if $y\langle t\rangle$ is (R,E)-separated then $y\langle t\rangle$ is (R,E)-preseparated

and we note that

(5) $y\langle t\rangle$ is (R,E)-separated $\Leftrightarrow$ $y\langle\hat{t}\rangle$ is (R,E)-separated

and we also note that

(6) if either $E\langle t\rangle$ is archimedian or $E\langle\hat{t}\rangle$ is archimedian or $y\langle t\rangle$ is R-ultraseparated or $y\langle\hat{t}\rangle$ is R-ultraseparated then: $y\langle t\rangle$ is (R,E)-separated.

We say that y is (R,E)-<u>preseparated</u> to mean that $y\langle 1\rangle$ is (R,E)-preseparated. We say that y is (R,E)-<u>separated</u> to mean that $y\langle 1\rangle$ is (R,E)-separated.

We say that $y\langle t\rangle$ is (R,E)-<u>regular</u> to mean that x is (R,D)-regular. We note that

(7) $\quad y\langle t\rangle$ is (R,E)-regular $\Leftrightarrow$ $y\langle\hat{t}\rangle$ is (R,E)-regular

and we also note that

(8) $\left\{\begin{array}{l}\text{if either } E\langle t\rangle \text{ is archimedian or } E\langle\hat{t}\rangle \text{ is archimedian} \\ \text{or } y\langle t\rangle \text{ is R-ultraseparated or } y\langle\hat{t}\rangle \text{ is R-ultraseparated} \\ \text{then: } y\langle t\rangle \text{ is (R,E)-regular} \Leftrightarrow y\langle\hat{t}\rangle \text{ is R-regular.}\end{array}\right.$

We say that y is (R,E)-<u>regular</u> to mean that $y\langle 1\rangle$ is (R,E)-regular.

§28. Weighted isobars and weighted initial forms

Let R be a ring. Let G' be a nonnegative ordered additive abelian semigroup and let Q' be the rational completion of G'. Let Y be given where

$$\begin{cases} \text{either } Y \text{ is an indeterminate string over } R \\ \text{or } Y \text{ is an indeterminate net over } R \end{cases}$$

and let E be given where

$$\begin{cases} E \in G'(o(Y)) \text{ in case } Y \text{ is a string} \\ \text{whereas } E \in G'(\ell(Y)) \text{ in case } Y \text{ is a net.} \end{cases}$$

For any $F \in R[Y]_Q$ we define

$$\mathrm{Ord}[R,Y,E](F) = \min \mathrm{inpo}(\mathrm{supt}(F),E)$$

and

$$\mathrm{Deg}[R,Y,E](F) = \max \mathrm{inpo}(\mathrm{supt}(F),E)$$

and we note that:

$$\begin{aligned} F \neq 0 &\Leftrightarrow \mathrm{Ord}[R,Y,E](F) \in \mathrm{inpo}(Q,E) \\ &\Leftrightarrow \mathrm{Deg}[R,Y,E](F) \in \mathrm{inpo}(Q,E) \end{aligned}$$

whereas:

$$\begin{aligned} F = 0 &\Leftrightarrow \mathrm{Ord}[R,Y,E](F) = \infty \\ &\Leftrightarrow \mathrm{Deg}[R,Y,E](F) = -\infty \end{aligned}$$

and moreover:

$$\text{if } 0 \neq F \in R[Y] \text{ then } \begin{cases} \mathrm{Ord}[R,Y,E](F) \in \mathrm{inpo}(Z,E) \\ \text{and} \\ \mathrm{Deg}[R,Y,E](F) \in \mathrm{inpo}(Z,E) \end{cases}$$

and we also observe that for any t where

$$\begin{cases} t \text{ is a string-restriction in case } Y \text{ is a string} \\ \text{whereas } t \text{ is a net-restriction in case } Y \text{ is a net} \end{cases}$$

we have that

$$\text{if } 0 \neq F \in R[Y\langle t\rangle]_Q \text{ then } \begin{cases} \mathrm{Ord}[R,Y,E](F) \in \mathrm{inpo}(Q,E\langle t\rangle) \\ \text{and} \\ \mathrm{Deg}[R,Y,E](F) \in \mathrm{inpo}(Q,E\langle t\rangle) \end{cases}$$

whereas:

$$\text{if } 0 \neq F \in R[Y\langle t\rangle] \text{ then } \begin{cases} \mathrm{Ord}[R,Y,E](F) \in \mathrm{inpo}(Z,E\langle t\rangle) \\ \text{and} \\ \mathrm{Deg}[R,Y,E](F) \in \mathrm{inpo}(Z,E\langle t\rangle). \end{cases}$$

For any $F' \subset R[Y]_Q$ we put

$$\mathrm{Ord}[R,Y,E](F') = \{\mathrm{Ord}[R,Y,E](F): F \in F'\}$$

and

$$\mathrm{Deg}[R,Y,E](F') = \{\mathrm{Deg}[R,Y,E](F): F \in F'\} .$$

If $G' = Z$ then for any $F' \subset R[Y]$ we put

$$\text{Ord}[R,Y,E]((F')) = \min \text{Ord}[R,Y,E](F')$$

and

$$\text{Deg}[R,Y,E]((F')) = \max \text{Ord}[R,Y,E](F').$$

For any $u \in Q'$ and $P \in \{=,\geq,>,<,\leq\}$ we define the

R-submodule $\text{Iso}(R,Y,EPu)_Q$ of $R[Y]_Q$

by putting

$$\text{Iso}(R,Y,EPu)_Q = \sum_{j\in Q(EPu)} Y_R^j$$

and we define the

R-submodule $\text{Iso}(R,Y,EPu)$ of $R[Y]$

by putting

$$\text{Iso}(R,Y,EPu) = \text{Iso}(R,Y,EPu)_Q \cap R[Y]$$

and we note that

$$\text{if } P \in \{\geq,>\} \text{ then } \begin{cases} \text{Iso}(R,Y,EPu)_Q \text{ is an ideal in } R[Y]_Q \\ \text{and} \\ \text{Iso}(R,Y,EPu) \text{ is an ideal in } R[Y]. \end{cases}$$

We observe that for any $u \in Q'$ we have

$$\text{Iso}(R,Y,E=u)_Q = \{F \in R[Y]_Q : \text{inpo}(j,E) = u \text{ for all } j \in \text{supt}(F)\}$$

$$\text{Iso}(R,Y,E \geq u)_Q = \{F \in R[Y]_Q : \text{Ord}[R,Y,E](F) \geq u\}$$

$$\mathrm{Iso}(R,Y,E>u)_Q = \{F \in R[Y]_Q\colon \mathrm{Ord}[R,Y,E](F) > u\}$$

$$\mathrm{Iso}(R,Y,E \le u)_Q = \{F \in R[Y]_Q\colon \mathrm{Deg}[R,Y,E](F) \le u\}$$

and

$$\mathrm{Iso}(R,Y,E<u)_Q = \{F \in R[Y]_Q\colon \mathrm{Deg}[R,Y,E](F) < u\} .$$

Given any $R_0 \subset R$ with $0 \in R_0$ and given any $u \in Q'$ and $P \in \{=,\ge,>,<,\le\}$, we put

$$\mathrm{Iso}(R_0,Y,EPu)_Q = \mathrm{Iso}(R,Y,EPu)_Q \cap R_0[Y]_Q$$

and

$$\mathrm{Iso}(R_0,Y,EPu) = \mathrm{Iso}(R,Y,EPu) \cap R_0[Y]$$

and for any t where

$$\begin{cases} t \text{ is a string-restriction in case } Y \text{ is a string} \\ \text{whereas } t \text{ is a net-restriction in case } Y \text{ is a net} \end{cases}$$

we put

$$\mathrm{Iso}(R_0,Y\langle t\rangle,EPu)_Q = \mathrm{Iso}(R,Y,EPu)_Q \cap R_0[Y\langle t\rangle]_Q$$

and

$$\mathrm{Iso}(R_0,Y\langle t\rangle,EPu) = \mathrm{Iso}(R,Y,EPu) \cap R_0[Y\langle t\rangle].$$

Given any ring-homomorphism $g\colon R \to R'$ and given any $u \in Q'$ and $P \in \{=,\ge,>,<,\le\}$, we define

$$Sub[g,Y,EPu]_Q\colon Iso(R,Y,EPu)_Q \to R'[Y]_Q$$

and

$$Sub[g,Y,EPu]\colon Iso(R,Y,EPu) \to R'[Y]$$

to be the g-homomorphisms induced by $Sub[g,Y]_Q$ and we define

$$Sub[g,Y,EPu]^*_Q\colon Iso(R,Y,EPu)_Q \to Iso(g(R),Y,EPu)_Q$$

and

$$Sub[g,Y,EPu]^*\colon Iso(R,Y,EPu) \to Iso(g(R),Y,EPu)$$

to be the g-epimorphisms induced by $Sub[g,Y]_Q$ and we observe that

$$\begin{aligned} ker(Sub[g,Y,EPu]_Q) &= ker(Sub[g,Y,EPu]^*_Q) \\ &= Iso(ker(g),Y,EPu)_Q \end{aligned}$$

and

$$\begin{aligned} ker(Sub[g,Y,EPu]) &= ker(Sub[g,Y,EPu]^*) \\ &= Iso(ker(g),Y,EPu) \end{aligned}$$

and now for any t where

$$\begin{cases} t \text{ is a string-restriction in case } Y \text{ is a string} \\ \text{whereas } t \text{ is a net-restriction in case } Y \text{ is a net} \end{cases}$$

we define

$$Sub[g,Y\langle t\rangle,EPu]_Q\colon Iso(R,Y\langle t\rangle,EPu)_Q \to R'[Y]_Q$$

and

$$\mathrm{Sub}[g,Y\langle t\rangle,\mathrm{EPu}]\colon \mathrm{Iso}(R,Y\langle t\rangle,\mathrm{EPu}) \to R'[Y]$$

to be the g-homomorphisms induced by $\mathrm{Sub}[g,Y]_Q$ and we define

$$\mathrm{Sub}[g,Y\langle t\rangle,\mathrm{EPu}]^*_Q\colon \mathrm{Iso}(R,Y\langle t\rangle,\mathrm{EPu})_Q \to \mathrm{Iso}(g(R),Y\langle t\rangle,\mathrm{EPu})_Q$$

and

$$\mathrm{Sub}[g,Y\langle t\rangle,\mathrm{EPu}]^*\colon \mathrm{Iso}(R,Y\langle t\rangle,\mathrm{EPu}) \to \mathrm{Iso}(g(R),Y\langle t\rangle,\mathrm{EPu})$$

to be the g-epimorphisms induced by $\mathrm{Sub}[g,Y]_Q$ and we observe that

$$\begin{aligned}\ker(\mathrm{Sub}[g,Y\langle t\rangle,\mathrm{EPu}]_Q) &= \ker(\mathrm{Sub}[g,Y\langle t\rangle,\mathrm{EPu}]^*_Q)\\ &= \mathrm{Iso}(\ker(g),Y\langle t\rangle,\mathrm{EPu})_Q\end{aligned}$$

and

$$\begin{aligned}\ker(\mathrm{Sub}[g,Y\langle t\rangle,\mathrm{EPu}]) &= \ker(\mathrm{Sub}[g,Y\langle t\rangle,\mathrm{EPu}]^*)\\ &= \mathrm{Iso}(\ker(g),Y\langle t\rangle,\mathrm{EPu}).\end{aligned}$$

Given any $u \in Q'$ and $P \in \{=,\geq,>,<,\leq\}$, we define the R-homomorphism

$$\mathrm{Iso}[R,Y,\mathrm{EPu}]_Q\colon R[Y]_Q \to R[Y]_Q$$

by putting

$$\mathrm{Iso}[R,Y,\mathrm{EPu}]_Q(F) = \sum_{j\in Q(\mathrm{EPu})} F[j]Y^j \quad \text{for all } F \in R[Y]_Q$$

and we define

$$\mathrm{Iso}[R,Y,EPu]: R[Y] \to R[Y]$$

to be the R-homomorphism induced by $\mathrm{Iso}[R,Y,EPu]_Q$ and we define

$$\mathrm{Iso}[R,Y,EPu]_Q^*: R[Y]_Q \to \mathrm{Iso}(R,Y,EPu)_Q$$

and

$$\mathrm{Iso}[R,Y,EPu]^*: R[Y] \to \mathrm{Iso}(R,Y,EPu)$$

to be the R-epimorphisms induced by $\mathrm{Iso}[R,Y,EPu]_Q$, and for any t where

$$\begin{cases} t \text{ is a string-restriction in case } Y \text{ is a string} \\ \text{whereas } t \text{ is a net-restriction in case } Y \text{ is a net} \end{cases}$$

we define

$$\mathrm{Iso}[R,Y\langle t\rangle,EPu]_Q: R[Y\langle t\rangle]_Q \to R[Y]_Q$$

and

$$\mathrm{Iso}[R,Y\langle t\rangle,EPu]: R[Y\langle t\rangle] \to R[Y]$$

to be the R-homomorphisms induced by $\mathrm{Iso}[R,Y,EPu]_Q$ and we define

$$\mathrm{Iso}[R,Y\langle t\rangle,EPu]_Q^*: R[Y\langle t\rangle]_Q \to \mathrm{Iso}(R,Y\langle t\rangle,EPu)_Q$$

and

$$\mathrm{Iso}[R,Y\langle t\rangle,EPu]^*: R[Y\langle t\rangle] \to \mathrm{Iso}(R,Y\langle t\rangle,EPu)$$

to be the R-epimorphisms induced by $\mathrm{Iso}[R,Y,EPu]_Q$.

Given any $u \in Q'$ we define

$$\mathrm{Info}[R,Y,E=u]_Q\colon \mathrm{Iso}(R,Y,E \geq u)_Q \to R[Y]_Q$$

and

$$\mathrm{Info}[R,Y,E=u]\colon \mathrm{Iso}(R,Y,E \geq u) \to R[Y]$$

to be the R-homomorphisms induced by $\mathrm{Iso}[R,Y,E=u]_Q$ and we define

$$\mathrm{Info}[R,Y,E=u]_Q^*\colon \mathrm{Iso}(R,Y,E \geq u)_Q \to \mathrm{Iso}(R,Y,E=u)_Q$$

and

$$\mathrm{Info}[R,Y,E=u]^*\colon \mathrm{Iso}(R,Y,E \geq u) \to \mathrm{Iso}(R,Y,E=u)$$

to be the R-epimorphisms induced by $\mathrm{Iso}[R,Y,E=u]_Q$ and we observe that

$$\begin{aligned}\ker(\mathrm{Info}[R,Y,E=u]_Q) &= \ker(\mathrm{Info}[R,Y,E=u]_Q^*)\\ &= \mathrm{Iso}(R,Y,E>u)_Q\end{aligned}$$

and

$$\begin{aligned}\ker(\mathrm{Info}[R,Y,E=u]) &= \ker(\mathrm{Info}[R,Y,E=u]^*)\\ &= \mathrm{Iso}(R,Y,E>u)\end{aligned}$$

and now for any t where

$$\begin{cases} t \text{ is a string-restriction in case } Y \text{ is a string}\\ \text{whereas } t \text{ is a net-restriction in case } Y \text{ is a net}\end{cases}$$

we define

$$\text{Info}[R,Y\langle t\rangle,E=u]_Q\colon \text{Iso}(R,Y\langle t\rangle,E\geq u)_Q \to R[Y]_Q$$

and

$$\text{Info}[R,Y\langle t\rangle,E=u]\colon \text{Iso}(R,Y\langle t\rangle,E\geq u) \to R[Y]$$

to be the R-homomorphisms induced by $\text{Iso}[R,Y,E=u]_Q$ and we define

$$\text{Info}[R,Y\langle t\rangle,E=u]_Q^*\colon \text{Iso}(R,Y\langle t\rangle,E\geq u)_Q \to \text{Iso}(R,Y\langle t\rangle,E=u)_Q$$

and

$$\text{Info}[R,Y\langle t\rangle,E=u]^*\colon \text{Iso}(R,Y\langle t\rangle,E\geq u) \to \text{Iso}(R,Y\langle t\rangle,E=u)$$

to be the R-epimorphisms induced by $\text{Iso}[R,Y,E=u]_Q$ and we observe that

$$\begin{aligned}\ker(\text{Info}[R,Y\langle t\rangle,E=u]_Q) &= \ker(\text{Info}[R,Y\langle t\rangle,E=u]_Q^*)\\ &= \text{Iso}(R,Y\langle t\rangle,E>u)_Q\end{aligned}$$

and

$$\begin{aligned}\ker(\text{Info}[R,Y\langle t\rangle,E=u]) &= \ker(\text{Info}[R,Y\langle t\rangle,E=u]^*)\\ &= \text{Iso}(R,Y\langle t\rangle,E>u).\end{aligned}$$

Given any ring-homomorphism $g\colon R\to R'$ and any $u\in Q'$ we define the g-epimorphism

$$\text{Info}[g,Y,E=u]_Q^*\colon \text{Iso}(R,Y,E\geq u)_Q \to \text{Iso}(g(R),Y,E=u)_Q$$

by considering the following diagram

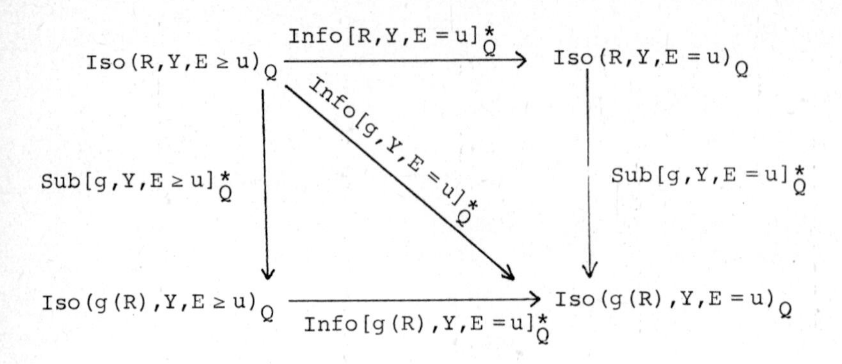

where the diagonal is yet to be defined; the undiagonaled rectangle is obviously commutative and hence the diagonal can be uniquely defined to make the diagonaled rectangle commutative.

Given any ring-homomorphism $g\colon R \to R'$ and any $u \in Q'$ we define the g-epimorphism

$$\text{Info}[g,Y,E=u]^*\colon \text{Iso}(R,Y,E \geq u) \to \text{Iso}(g(R),Y,E=u)$$

by everywhere deleting the subscript Q in the above paragraph; alternatively, we can define it to be the g-epimorphism induced by $\text{Info}[g,Y,E=u]^*_Q$.

Given any ring-homomorphism $g\colon R \to R'$ and any $u \in Q'$ we define

$$\text{Info}[g,Y,E=u]_Q\colon \text{Iso}(R,Y,E \geq u)_Q \to R'[Y]_Q$$

to be the g-homomorphism induced by $\text{Info}[g,Y,E=u]^*_Q$ which means that it is the composition

$$\text{Iso}(R,Y,E \geq u)_Q \xrightarrow{\text{Info}[g,Y,E=u]^*_Q} \text{Iso}(g(R),Y,E=u)_Q \to R'[Y]_Q$$

where the second arrow is the natural injection, and we define

$$\text{Info}[g,Y,E=u]: \text{Iso}(R,Y,E \geq u) \to R'[Y]$$

to be the g-homomorphism induced by $\text{Info}[g,Y,E=u]^*$ which means that it is the composition

$$\text{Iso}(R,Y,E \geq u)_Q \xrightarrow{\text{Info}[g,Y,E=u]^*} \text{Iso}(g(R),Y,E=u) \to R'[Y]$$

where again the second arrow is the natural injection.

Given any ring-homomorphism $g: R \to R'$ and given any $u \in Q'$ and any t where

$$\begin{cases} t \text{ is a string-restriction in case } Y \text{ is a string} \\ \text{whereas } t \text{ is a net-restriction in case } Y \text{ is a net} \end{cases}$$

we define

$$\text{Info}[g,Y\langle t\rangle,E=u]_Q: \text{Iso}(R,Y\langle t\rangle,E \geq u)_Q \to R'[Y]_Q$$

and

$$\text{Info}[g,Y\langle t\rangle,E=u]: \text{Iso}(R,Y\langle t\rangle,E \geq u) \to R'[Y]$$

to be the g-homomorphisms induced by $\text{Info}[g,Y,E=u]_Q$ and we define

$$\text{Info}[g,Y\langle t\rangle,E=u]_Q^*: \text{Iso}(R,Y\langle t\rangle,E \geq u)_Q \to \text{Iso}(g(R),Y\langle t\rangle,E=u)_Q$$

and

$$\text{Info}[g,Y\langle t\rangle,E=u]^*: \text{Iso}(R,Y\langle t\rangle,E\geq u) \to \text{Iso}(g(R),Y\langle t\rangle,E=u)$$

to be the R-epimorphisms induced by $\text{Info}[R,Y,E=u]_Q$; we note that, upon changing Y to $Y\langle t\rangle$ everywhere in the above diagram, we get a commutative diagonaled rectangle; we also note that, upon changing Y to $Y\langle t\rangle$ everywhere and deleting the subscript Q everywhere in the above diagram, we again get a commutative diagonaled rectangle.

Given any $u \in Q'$ and given any $P \in \{=,\geq,>,<,\leq\}$ and any y where

$$\begin{cases} y \in R(o(Y)) & \text{in case } Y \text{ is a string} \\ \text{whereas } y \in R(\ell(Y)) & \text{in case } Y \text{ is a net} \end{cases}$$

we define

$$\text{sub}[R,Y=y,EPu]: \text{Iso}(R,Y,EPu) \to R$$

to be the R-homomorphism induced by $\text{sub}[R,Y=y]$ and we put

$$\text{iso}(R,y,EPu) = \text{sub}[R,Y=y](\text{Iso}(R,Y,EPu))$$

and we define

$$\text{sub}[R,Y=y,EPu]^*: \text{Iso}(R,Y,EPu) \to \text{iso}(R,y,EPu)$$

to be the R-epimorphism induced by $\text{sub}[R,Y=y]$, and for any t where

$$\begin{cases} t \text{ is a string-restriction in case } Y \text{ is a string} \\ \text{whereas } t \text{ is a net-restriction in case } Y \text{ is a net} \end{cases}$$

we define

$$sub[R, Y\langle t\rangle = y, EPu]: Iso(R, Y\langle t\rangle, EPu) \rightarrow R$$

to be the R-homomorphism induced by $sub[R, Y = y]$ and we put

$$iso(R, y\langle t\rangle, EPu) = sub[R, Y = y](Iso(R, Y\langle t\rangle, EPu))$$

and we define

$$sub[R, Y\langle t\rangle = y, EPu]^{*}: Iso(R, Y\langle t\rangle, EPu) \rightarrow iso(R, y\langle t\rangle, EPu)$$

to be the R-epimorphism induced by $sub[R, Y = y]$.

§29. Initial forms for regular strings

Let R be a ring. Let x be an R-string and let X be an indeterminate string over R with $o(X) = o(x)$. Let G' be a nonnegative ordered additive abelian semigroup and let Q' be the rational completion of G'. Let $D \in G'(o(x))$. Let t be a string-restriction and let

$$\hat{t} = \mathrm{supt}(o(x),t) \cap \mathrm{supt}(D).$$

LEMMA 1. For any $w \in Q'$ and $P \in \{=,\geq,>,<,\leq\}$ we have

$$\mathrm{iso}(R,x\langle t\rangle,DPw) = \mathrm{iso}(R,x\langle \hat{t}\rangle,DPw).$$

PROOF. Obvious.

LEMMA 2. Let

$$t' = \{c' \in \hat{t}\colon \text{for every } c \in \hat{t} \text{ we have } n(c,c')D(c') \geq D(c) \text{ for some } n(c,c') \in Z\}.$$

Then

$$\bigcap_{n=1}^{\infty} x\langle t'\rangle_R^n = \bigcap_{w \in \mathrm{inpo}(Z,D\langle t'\rangle)} \mathrm{iso}(R,x\langle t'\rangle,D > w)$$

$$= \bigcap_{w \in \mathrm{inpo}(Z,D\langle t\rangle)} \mathrm{iso}(R,x\langle t\rangle,D > w).$$

Moreover, if $\hat{t} \neq \emptyset$ then: $t' \neq \emptyset$ and

$$\bigcap_{n=1}^{\infty} x\langle t'\rangle_R^n = \bigcap_{w \in \text{inpo}(Z,D\langle t'\rangle)} \text{iso}(R,x\langle t'\rangle,D \geq w)$$

$$= \bigcap_{w \in \text{inpo}(Z,D\langle t\rangle)} \text{iso}(R,x\langle t\rangle,D \geq w).$$

PROOF. If $\hat{t} = \emptyset$ then each one of the first three intersections is reduced to $\{0\}$. So now suppose that $\hat{t} \neq \emptyset$. Then $t' \neq \emptyset$ and we can find

(1) $c_1 \in t'$ and $c_2 \in t'$

and

(2) $0 \neq A \in Z$

such that

(3) $AD(c_2) \geq D(c_1) \geq D(c) \geq D(c_2)$ for all $c \in t'$.

By (3) it follows that

(4) $AD(c_2)\text{abs}(i) \geq \text{inpo}(i,D) \geq D(c_2)\text{abs}(i)$ for all $i \in Z(o(x),t')$.

By (1) we have

(5) $D(c_2) > 0$

and

(6) $D(c_2) > nD(c)$ for all $n \in Z$ and $c \in t \backslash t'$.

If $n = 0$ then obviously, whereas if $0 \neq n \in Z$ then by (2), (4) and (5), we see that

$$\mathrm{iso}(R,x\langle t'\rangle, D \geq AnD(c_2)) \subset x\langle t'\rangle_R^n \subset \mathrm{iso}(R,x\langle t'\rangle, D \geq nD(c_2))$$

and hence upon taking intersections as n varies over Z we get

$$(7)\qquad \begin{cases} \bigcap\limits_{n\in Z} \mathrm{iso}(R,x\langle t'\rangle, D \geq AnD(c_2)) \\ \subset \bigcap\limits_{n\in Z} x\langle t'\rangle_R^n \\ \subset \bigcap\limits_{n\in Z} \mathrm{iso}(R,x\langle t'\rangle, D \geq nD(c_2)). \end{cases}$$

For any $P \in \{\geq,>\}$, because $Z \supset AZ$, we have

$$\bigcap_{n\in Z} \mathrm{iso}(R,x\langle t'\rangle, DPnD(c_2)) \subset \bigcap_{n\in AZ} \mathrm{iso}(R,x\langle t'\rangle, DPnD(c_2))$$

and obviously we have

$$\bigcap_{n\in AZ} \mathrm{iso}(R,x\langle t'\rangle, DPnD(c_2)) = \bigcap_{n\in Z} \mathrm{iso}(R,x\langle t'\rangle, DPAnD(c_2))$$

and so we conclude that

$$(8)\qquad \begin{cases} \text{for any } P \in \{\geq,>\} \text{ we have} \\ \bigcap\limits_{n\in Z} \mathrm{iso}(R,x\langle t'\rangle, DPnD(c_2)) \subset \bigcap\limits_{n\in Z} \mathrm{iso}(R,x\langle t'\rangle, DPAnD(c_2)). \end{cases}$$

Now obviously

$$\bigcap_{n\in Z} x\langle t'\rangle_R^n = \bigcap_{n=1}^{\infty} x\langle t'\rangle_R^n$$

and hence by (7) and (8) we see that

$$(9)\quad \begin{cases} \bigcap_{n\in Z} \text{iso}(R,x\langle t'\rangle,D \geq AnD(c_2)) \\ = \bigcap_{n=1}^{\infty} x\langle t'\rangle_R^n \\ = \bigcap_{n\in Z} \text{iso}(R,x\langle t'\rangle,D \geq nD(c_2)). \end{cases}$$

By (4) we see that

$$\begin{cases} \text{for any } P \in \{\geq,>\} \text{ and } n \in Z \text{ we have} \\ \text{iso}(R,x\langle t'\rangle,DPAnD(c_2)) \\ \subset \bigcap_{i\in Z(o(x)=n,t')} \text{iso}(R,x\langle t'\rangle,DPinpo(i,D)) \\ \subset \text{iso}(R,x\langle t'\rangle,DPnD(c_2)) \end{cases}$$

and hence, upon taking intersections as n varies over Z, we see that

$$(10)\quad \begin{cases} \text{for any } P \in \{\geq,>\} \text{ we have} \\ \bigcap_{n\in Z} \text{iso}(R,x\langle t'\rangle,DPAnD(c_2)) \\ \subset \bigcap_{n\in Z} \bigcap_{i\in Z(o(x)=n,t')} \text{iso}(R,x\langle t'\rangle,DPinpo(i,D)) \\ \subset \bigcap_{n\in Z} \text{iso}(R,x\langle t'\rangle,DPnD(c_2)) . \end{cases}$$

Clearly

$$(11)\quad \begin{cases} \text{for any } P \in \{\geq, >\} \text{ we have} \\ \bigcap_{n\in Z} \bigcap_{i\in Z(o(x)=n,t')} \text{iso}(R, x\langle t'\rangle, DPinpo(i,D)) \\ = \bigcap_{w\in inpo(Z,D\langle t'\rangle)} \text{iso}(R, x\langle t'\rangle, DPw). \end{cases}$$

By (8), (10) and (11) we see that

$$(12)\quad \begin{cases} \text{for any } P \in \{\geq, >\} \text{ we have} \\ \bigcap_{n\in Z} \text{iso}(R, x\langle t'\rangle, DPnD(c_2)) \\ = \bigcap_{w\in inpo(Z,D\langle t'\rangle)} \text{iso}(R, x\langle t'\rangle, DPw) \\ = \bigcap_{n\in Z} \text{iso}(R, x\langle t'\rangle, DPAnD(c_2)). \end{cases}$$

In view of (5), for every $n \in Z$ we have

$$\text{iso}(R, x\langle t'\rangle, D \geq (n+1)D(c_2)) \subset \text{iso}(R, x\langle t'\rangle, D > nD(c_2))$$

and obviously for every $n \in Z$ we have

$$\text{iso}(R, x\langle t'\rangle, D > nD(c_2)) \subset \text{iso}(R, x\langle t'\rangle, D \geq nD(c_2))$$

and hence upon taking intersections as n various Z we get

$$
(13)\quad \begin{cases} \bigcap_{n\in Z} \mathrm{iso}(R,x\langle t'\rangle,D \geq (n+1)D(c_2)) \\ \subset \bigcap_{n\in Z} \mathrm{iso}(R,x\langle t'\rangle,D > nD(c_2)) \\ \subset \bigcap_{n\in Z} \mathrm{iso}(R,x\langle t'\rangle,D \geq nD(c_2)) \;. \end{cases}
$$

Clearly

$$
(14)\quad \bigcap_{n\in Z} \mathrm{iso}(R,x\langle t'\rangle,D \geq (n+1)D(c_2)) = \bigcap_{0\neq n\in Z} \mathrm{iso}(R,x\langle t'\rangle,D \geq nD(c_2)).
$$

Obviously

$$
\bigcap_{0\neq n\in Z} \mathrm{iso}(R,x\langle t'\rangle,D \geq nD(c_2)) \subset \bigcap_{n\in\{0\}} \mathrm{iso}(R,x\langle t'\rangle,D \geq nD(c_2))
$$

and hence

$$
(15)\quad \bigcap_{0\neq n\in Z} \mathrm{iso}(R,x\langle t'\rangle,D \geq nD(c_2)) = \bigcap_{n\in Z} \mathrm{iso}(R,x\langle t'\rangle,D \geq nD(c_2)).
$$

By (14) and (15) we get that

$$
(16)\quad \bigcap_{n\in Z} \mathrm{iso}(R,x\langle t'\rangle,D \geq (n+1)D(c_2)) = \bigcap_{n\in Z} \mathrm{iso}(R,x\langle t'\rangle,D \geq nD(c_2)).
$$

By (13) and (16) we see that

$$
(17)\quad \bigcap_{n\in Z} \mathrm{iso}(R,x\langle t'\rangle,D \geq nD(c_2)) = \bigcap_{n\in Z} \mathrm{iso}(R,x\langle t'\rangle,D > nD(c_2)).
$$

Clearly

$$(18)\begin{cases}\text{for any } P \in \{\geq,>\} \text{ we have} \\ \bigcap_{n\in Z} \mathrm{iso}(R,x\langle t'\rangle,DPA(n+2)D(c_2)) = \bigcap_{2\leq n\in Z} \mathrm{iso}(R,x\langle t'\rangle,DPAnD(c_2)).\end{cases}$$

Obviously for any $P \in \{\geq,>\}$ we have

$$\bigcap_{2\leq n\in Z} \mathrm{iso}(R,x\langle t'\rangle,DPAnD(c_2)) \subset \bigcap_{n\in[0,1]} \mathrm{iso}(R,x\langle t'\rangle,DPAnD(c_2))$$

and hence

$$(19)\begin{cases}\text{for any } P \in \{\geq,>\} \text{ we have} \\ \bigcap_{2\leq n\in Z} \mathrm{iso}(R,x\langle t'\rangle,DPAnD(c_2)) = \bigcap_{n\in Z} \mathrm{iso}(R,x\langle t'\rangle,DPAnD(c_2)).\end{cases}$$

By (18) and (19) we see that

$$(20)\begin{cases}\text{for any } P \in \{\geq,>\} \text{ we have} \\ \bigcap_{n\in Z} \mathrm{iso}(R,x\langle t'\rangle,DPA(n+2)D(c_2)) = \bigcap_{n\in Z} \mathrm{iso}(R,x\langle t'\rangle,DPAnD(c_2)).\end{cases}$$

Upon letting

$$(21)\qquad t'' = t\backslash t'$$

by (1), (3), (5) and (6) we see that for any $P \in \{\geq,>\}$ we have

$$\begin{cases}\bigcap_{n\in Z} \bigcap_{\substack{i\in Z(o(x)=n+1,t') \\ j\in Z(o(x),t'')}} \mathrm{iso}(R,x\langle t\rangle,DPinpo(i+j,D)) \\ \subset \bigcap_{j\in Z(o(x),t'')} \mathrm{iso}(R,x\langle t\rangle,DPinpo(j,D))\end{cases}$$

and hence

(22)
$$\left\{\begin{array}{l}\text{for any } P \in \{\geq,>\} \text{ we have} \\ \bigcap_{n\in Z} \bigcap_{\substack{i\in Z(o(x)=n+1,t') \\ j\in Z(o(x),t'')}} iso(R,x\langle t\rangle,DPinpo(i+j,D)) \\ = \bigcap_{n\in Z} \bigcap_{\substack{i\in Z(o(x)=n,t') \\ j\in Z(o(x),t'')}} iso(R,x\langle t\rangle,DPinpo(i+j,D)).\end{array}\right.$$

Obviously

(23)
$$\left\{\begin{array}{l}\text{for any } P \in \{\geq,>\} \text{ we have} \\ \bigcap_{n\in Z} \bigcap_{\substack{i\in Z(o(x)=n,t') \\ j\in Z(o(x),t'')}} iso(R,x\langle t\rangle,DPinpo(i+j,D)) \\ = \bigcap_{w\in inpo(Z,D\langle t\rangle)} iso(R,x\langle t\rangle,DPw).\end{array}\right.$$

By (22) and (23) we see that

(24)
$$\left\{\begin{array}{l}\text{for any } P \in \{\geq,>\} \text{ we have} \\ \bigcap_{n\in Z} \bigcap_{\substack{i\in Z(o(x)=n+1,t') \\ j\in Z(o(x),t'')}} iso(R,x\langle t\rangle,DPinpo(i+j,D)) \\ = \bigcap_{w\in inpo(Z,D\langle t\rangle)} iso(R,x\langle t\rangle,DPw).\end{array}\right.$$

By (2), (4), (6) and (21) we see that

$$(25)\quad \begin{cases} \text{for any } P \in \{\geq,>\} \text{ and } n \in Z \text{ we have} \\ \mathrm{iso}(R,x\langle t'\rangle,\mathrm{DPA}(n+2)\mathrm{D}(c_2)) \\ \subset \bigcap\limits_{\substack{i\in Z(o(x)=n+1,t') \\ j\in Z(o(x),t'')}} \mathrm{iso}(R,x\langle t\rangle,\mathrm{DPinpo}(i+j,D)) \\ \subset \mathrm{iso}(R,x\langle t'\rangle,\mathrm{DPnD}(c_2)). \end{cases}$$

Upon taking intersections as n varies over Z, by (25) we see that

$$(26)\quad \begin{cases} \text{for any } P \in \{\geq,>\} \text{ we have} \\ \bigcap\limits_{n\in Z} \mathrm{iso}(R,x\langle t'\rangle,\mathrm{DPA}(n+2)\mathrm{D}(c_2)) \\ \subset \bigcap\limits_{n\in Z} \bigcap\limits_{\substack{i\in Z(o(x)=n+1,t') \\ j\in Z(o(x),t'')}} \mathrm{iso}(R,x\langle t\rangle,\mathrm{DPinpo}(i+j,D)) \\ \subset \bigcap\limits_{n\in Z} \mathrm{iso}(R,x\langle t'\rangle,\mathrm{DPnD}(c_2)). \end{cases}$$

By (8), (20), (24) and (26) we see that

$$(27)\quad \begin{cases} \text{for any } P \in \{\geq,>\} \text{ we have} \\ \bigcap\limits_{n\in Z} \mathrm{iso}(R,x\langle t'\rangle,\mathrm{DPAnD}(c_2)) \\ = \bigcap\limits_{w\in \mathrm{inpo}(Z,D\langle t\rangle)} \mathrm{iso}(R,x\langle t\rangle,\mathrm{DPw}) \\ = \bigcap\limits_{n\in Z} \mathrm{iso}(R,x\langle t'\rangle,\mathrm{DPnD}(c_2)). \end{cases}$$

The assertions of the Lemma now follow from (9), (12), (17) and (27).

LEMMA 3. Assume that $x\langle t\rangle$ is (R,D)-separated. Then given any

$$f \in R \setminus \bigcap_{w \in \mathrm{inpo}(Z, D\langle t\rangle)} \mathrm{iso}(R, x\langle t\rangle, D > w)$$

there exist $u \in \mathrm{inpo}(Z, D\langle \hat{t}\rangle)$ and $F \in \mathrm{Iso}(R, X\langle \hat{t}\rangle, D \geq u)$ such that $\mathrm{sub}[R, X = x](F) = f$ and $\mathrm{Info}[\mathrm{res}[R, x\langle \hat{t}\rangle], X, D = u](F) \neq 0$.

PROOF. We shall make induction on $\mathrm{card}(\hat{t})$. If $\mathrm{card}(\hat{t}) = 0$ then, in view of Lemma 1, it suffices to take $u = 0$ and $F = f$. So now suppose that $\mathrm{card}(\hat{t}) \neq 0$ and assume that the assertion is true for all values of $\mathrm{card}(\hat{t})$ smaller than the given one.

We can find

$$c_1 \in \hat{t} \quad \text{such that} \quad D(c_1) \geq D(c) \quad \text{for all} \quad c \in t.$$

Let

$$t'' = \{c \in \hat{t} \colon D(c_1) > nD(c\ ) \text{ for all } n \in Z\} \text{ and } t' = \hat{t} \setminus t''.$$

We can now find

$$c_2 \in t' \quad \text{such that} \quad D(c) \geq D(c_2) \quad \text{for all } c \in t'$$

and then we can find

$$0 \neq A \in Z \quad \text{such that} \quad AD(c_2) \geq D(c_1).$$

Thus we now have

(1) $\quad t' \cup t'' = \hat{t}$ and $t' \neq \emptyset = t' \cap t''$

(2) $\quad D(c') > nD(c'')$ whenever $n \in Z$, $c' \in t'$ and $c'' \in t''$

and

(3) $\quad \begin{cases} c_1 \in t' \text{ and } c_2 \in t' \text{ and } 0 \neq A \in Z \text{ with} \\ AD(c_2) \geq D(c_1) \geq D(c) \geq D(c_2) \text{ for all } c \in t' . \end{cases}$

Upon letting

(4) $\quad V = \text{ord}[R,x\langle t'\rangle](f)$

in view of Lemma 2 we have that

(5) $\quad V \in Z$

and there exists

$$H^* \in \text{Iso}(R,X\langle t'\rangle,=V) \text{ such that } \text{sub}[R,X=x](H^*) = f.$$

Let

(6) $\quad W = \{h \in \text{Iso}(R,X\langle t'\rangle, \leq 1+VA)\setminus\{0\}: \text{sub}[R,X=x](h) = f\}.$

Then $H^* \in W$; consequently $W \neq \emptyset$ and hence $\text{Ord}[R,X,D](W) \neq \emptyset$; also clearly

$$\text{Ord}[R,X,D](W) \subset \text{Ord}[R,X,D](\text{Iso}(R,X\langle t'\rangle, \leq 1+AV)\setminus\{0\}) = \begin{cases} \text{a finite} \\ \text{subset} \\ \text{of } G'; \end{cases}$$

therefore $\text{Ord}[R,X,D](W)$ is a nonempty finite subset of G' and hence upon letting

(7) $$u' = \max \mathrm{Ord}[R,X,D](W)$$

we have that

(8) $$u' \in G'$$

and there exists $H \in W$ such that

(9) $$\mathrm{Ord}[R,X,D](H) = u' .$$

Since $H \in W$, we have

(10) $$0 \neq H \in R[X\langle t'\rangle]$$

(11) $$\mathrm{sub}[R,X=x](H) = f$$

and

(12) $$\mathrm{Deg}[R,X](H) \leq 1 + VA .$$

Let

(13) $$I = \{i \in \mathrm{supt}(H): \mathrm{inpo}(i,D) = u'\}$$

(14) $$J = \{j \in \mathrm{supt}(H): H[j] \notin x\langle t'\rangle_R^1\}$$

and

(15) $$K = \{k \in \mathrm{supt}(H): H[k] \in x\langle t'\rangle_R^1\}.$$

By (4), (5), (10) and (11) there exists

(16) $$i_0 \in \mathrm{supt}(H)$$

such that

(17) $$\mathrm{abs}(i_0) \leq V\ .$$

For every $i \in I$ we have

$$\begin{aligned}
\mathrm{abs}(i)D(c_2) &= \sum_{c \in t'} i(c)D(c_2) && \text{by (10) and (13)} \\
&\leq \sum_{c \in t'} i(c)D(c) && \text{by (3)} \\
&= \mathrm{Ord}[R,X,D](H) && \text{by (10) and (13)} \\
&\leq \mathrm{inpo}(i_0,D) && \text{by (16)} \\
&= \sum_{c \in t'} i_0(c)D(c) && \text{by (10) and (16)} \\
&\leq \sum_{c \in t'} i_0(c)AD(c_2) && \text{by (3)} \\
&\leq VAD(c_2) && \text{by (10),(16) and (17).}
\end{aligned}$$

Therefore

(18) $$\mathrm{abs}(i) \leq VA \quad \text{for all} \quad i \in I\ .$$

By (15) we can write

(19) $$H[k] = \sum_{c \in t'} z(k,c)x(c) \quad \text{with} \quad z(k,c) \in R \quad \text{for all} \quad k \in K.$$

Upon letting

(20) $$H' = \sum_{i \in I \cap J} H[i]X^i$$

and

$$H'' = \sum_{i \in \mathrm{supt}(H) \setminus I} H[i]X^i + \sum_{\substack{k \in I \cap K \\ c \in t'}} z(k,c)X(c)X^k$$

by (6), (8) to (15), and (18) and (19), we see that

(21) $$H' + H'' \in W$$

and

(22) $$\mathrm{Ord}[R,X,D](H') \geq u' < \mathrm{Ord}[R,X,D](H'').$$

By (7) and (21) we get

(23) $$\mathrm{Ord}[R,X,D](H' + H'') \leq u' .$$

By (22) and (23) we get

(24) $$\mathrm{Ord}[R,X,D](H') = u'.$$

By (8), (20) and (24) we conclude that

(25) $$I \cap J \neq \emptyset .$$

[We note that if $t'' = \emptyset$ then, in view of (1), (9), (10), (11), (13), (14) and (25), it suffices to take $u = u'$ and

$F = H$. In other words, in case of $t" = \emptyset$ the proof is now finished without having invoked the induction hypothesis].

By (1) we have $\mathrm{card}(t") < \mathrm{card}(\hat{t})$ and hence, in view of Lemma 1 and (1) and (14), by the induction hypothesis, for every $j \in J$ there exists

$$(26) \qquad u_j \in \mathrm{inpo}(Z, D\langle t"\rangle)$$

and

$$(27) \qquad H_j \in \mathrm{Iso}(\mathrm{res}(R, x\langle t'\rangle), X\langle t"\rangle, D \geq u_j)$$

such that

$$(28) \qquad \mathrm{sub}[\mathrm{res}(R, x\langle t'\rangle), X = \mathrm{res}[R, x\langle t'\rangle](x)](H_j) = \mathrm{res}[R, x\langle t'\rangle](H[j])$$

and

$$(29) \qquad \mathrm{Info}[\mathrm{res}[(R, x\langle t\rangle), x\langle t'\rangle], X, D = u_j](H_j) \neq 0.$$

For every $j \in J$ we can find

$$F_{j,i} \in R \text{ with } \mathrm{res}[R, x\langle t'\rangle](F_{j,i}) = H_j[i] \text{ for all } i \in \mathrm{supt}(H_j)$$

and then upon letting

$$F_j = \sum_{i \in \mathrm{supt}(H_j)} F_{j,i} X^i$$

in view of (27) we see that

(30) $$F_j \in R[X\langle t''\rangle]$$

(31) $$\mathrm{supt}(F_j) = \mathrm{supt}(H_j)$$

and

(32) $$\mathrm{res}[R,x\langle t'\rangle](F_j[i]) = H_j[i] \quad \text{for all} \quad i \in \mathrm{supt}(H_j)$$

and now in view of (28) we see that

$$H[j] - \sum_{i\in \mathrm{supt}(F_j)} F_j[i]x^i \in x\langle t'\rangle_R^1$$

and hence we can write

(33) $$H[j] - \sum_{i\in \mathrm{supt}(F_j)} F_j[i]x^i = \sum_{c\in t'} z(j,c)x(c) \text{ with } z(j,c) \in R.$$

Upon letting

(34) $$F' = \sum_{j\in J} F_j X^j$$

(35) $$F'' = \sum_{\substack{k\in J\cup K \\ c\in t'}} z(k,c)X(c)X^k$$

and

(36) $$F = F' + F''$$

by (1), (10), (11), (14), (15), (19), (30), (31), (32) and (33) we see that

(37) $$F \in R[X\langle \hat{t}\rangle]$$

and

$$\text{sub}[R, X = x](F) = f. \tag{38}$$

By (14) and (25) we know that J is a nonempty finite set and hence upon letting

$$u = \min_{j \in J} (\text{inpo}(j, D) + u_j) \tag{39}$$

by (1), (10), (14), (26), (27), (29), (30), (31), (32) and (34) we see that

$$u \in \text{inpo}(Z, D\langle \hat{t} \rangle) \tag{40}$$

$$\text{Ord}[R, X, D](F') = u \tag{41}$$

and

$$\text{Info}[\text{res}[R, x\langle \hat{t} \rangle], X, D = u](F') \neq 0. \tag{42}$$

By (25) we can take some $j' \in I \cap J$ and then by (39) we have

$$u \leq \text{inpo}(j', D) + u_{j'}$$

and by (13) we have

$$\text{inpo}(j', D) = u'$$

and so in view of (26) we get

$$u \leq u' + u_{j'} \quad \text{with} \quad u_{j'} \in \text{inpo}(Z, D\langle t'' \rangle). \tag{43}$$

By (9), (14) and (15) we get

$$(44) \qquad u' \leq \mathrm{inpo}(k,D) \quad \text{for all} \quad k \in J \cup K.$$

By (2), (35), (43) and (44) we see that

$$(45) \qquad \mathrm{Ord}[R,X,D](F'') > u.$$

By (36), (37), (41), (42) and (45) we conclude that

$$F \in \mathrm{Iso}(R,X\langle \hat{t}\rangle ,D \geq u)$$

and

$$\mathrm{Info}[\mathrm{res}[R,x\langle \hat{t}\rangle],X,D = u](F) \neq 0$$

and in view of (38) this completes the proof.

LEMMA 4. Assume that $x\langle t\rangle$ is (R,D)-regular. Let $u \in Q'$ and

$$F_1 \in \mathrm{Iso}(R,X\langle \hat{t}\rangle ,D \geq u) \text{ and } F_2 \in \mathrm{Iso}(R,X\langle \hat{t}\rangle ,D \geq u)$$

be such that

$$\mathrm{Info}[\mathrm{res}[R,x\langle \hat{t}\rangle],X,D=u](F_1) \neq \mathrm{Info}[\mathrm{res}[R,x\langle \hat{t}\rangle ,X,D=u](F_2).$$

Then

$$\mathrm{sub}[R,X = x](F_1) \neq \mathrm{sub}[R,X = x](F_2).$$

PROOF. We shall make induction on $\mathrm{card}(\hat{t})$. The assertion is obvious when $\mathrm{card}(\hat{t}) = 0$. So now suppose that $\mathrm{card}(\hat{t}) > 0$

and assume that the assertion is true for all values of $\mathrm{card}(\hat{t})$ smaller than the given one. Let

$$c = \max \hat{t} \quad \text{and} \quad t' = \hat{t} \setminus \{c\}.$$

By assumption $x\langle \hat{t} \rangle$ is R-regular and hence

(1) for every $z \in R \setminus x\langle t' \rangle_R^1$ we have $zx(c) \notin x\langle t' \rangle_R^1$.

Upon letting

(2) $$F = F_1 - F_2$$

we have

$$\begin{cases} F \in \mathrm{Iso}(R, X\langle \hat{t} \rangle, D \geq u) \quad \text{and} \\ \mathrm{Info}[\mathrm{res}[R, x\langle \hat{t} \rangle], X, D = u](F) \neq 0 \end{cases}$$

and so upon letting

$$a = \mathrm{Deg}[R, X](F\langle c \rangle)$$

and

$$b = \mathrm{Deg}[\mathrm{res}(R, x\langle \hat{t} \rangle), X](\mathrm{Info}[\mathrm{res}[R, x\langle \hat{t} \rangle], X, D = u](F)\langle c \rangle)$$

we have

$$a \in Z \quad \text{and} \quad b \in [0, a]$$

and upon letting

$$v = u - bD(c)$$

$$H_0 = \sum_{n \in [0,b-1]} F\langle c \rangle [n] x(c)^n$$

$$H_1 = F\langle c \rangle [b]$$

$$H_2 = \sum_{n \in [b+1,a]} F\langle c \rangle [n] x(c)^{n-b-1}$$

we have

(3) $$v \in \mathrm{inpo}(Z, D\langle t' \rangle)$$

(4) $$H_i \in R[X\langle t' \rangle] \quad \text{for} \quad 0 \leq i \leq 2$$

(5) $$\mathrm{sub}[R, X = x](F) = \mathrm{sub}[R, X = x](H_0 + H_1 x(c)^b + H_2 x(c)^{b+1})$$

(6) $$H_0 \in \mathrm{Iso}(R, X\langle t' \rangle, D > v)$$

(7) $$H_1 \in \mathrm{Iso}(R, X\langle t' \rangle, D \geq v)$$

and

(8) $$\mathrm{Info}[\mathrm{res}[R, x\langle \hat{t} \rangle], X, D = v](H_1) \neq 0 .$$

We shall now divide the rest of the argument into two cases according as

$$\begin{cases} \mathrm{sub}[R, X = x](H_2) \in \mathrm{iso}(R, x\langle t' \rangle, D \geq v) \\ \text{or} \\ \mathrm{sub}[R, X = x](H_2) \notin \mathrm{iso}(R, x\langle t' \rangle, D \geq v). \end{cases}$$

First consider the case when

$$\text{(11)} \qquad \text{sub}[R, X=x](H_2) \in \text{iso}(R, x\langle t'\rangle, D \geq v).$$

Now there exists

$$\text{(12)} \qquad \bar{H}_2 \in \text{Iso}(R, X\langle t'\rangle, D \geq v)$$

such that

$$\text{(13)} \qquad \text{sub}[R, X=x](\bar{H}_2) = \text{sub}[R, X=x](H_2).$$

Upon letting

$$H = H_1 + \bar{H}_2 x(c)$$

by (7) and (12) we have

$$\text{(14)} \qquad H \in \text{Iso}(R, X\langle t'\rangle, D \geq v)$$

and by (5) and (13) we have

$$\text{(15)} \qquad \text{sub}[R, X=x](F) = \text{sub}[R, X=x](H_0 + Hx(c)^b)$$

and obviously we have

$$\text{(16)} \qquad \text{Info}[\text{res}[R, x\langle\hat{t}\rangle], X, D=v](H) = \text{Info}[\text{res}[R, x\langle\hat{t}\rangle], X, D=v](H_1).$$

Now by (8) and (16)

$$\text{Info}[\text{res}[R, x\langle\hat{t}\rangle], X, D=v](H) \neq 0$$

and hence a fortiori

(17) $$\mathrm{Info}[\mathrm{res}[R,x\langle t'\rangle],X,D=v](H) \neq 0.$$

By (1), (14) and (17) we see that

$$\begin{cases} Hx(c)^b \in \mathrm{Iso}(R,X\langle t'\rangle,D\geq v) \quad \text{and} \\ \mathrm{Info}[\mathrm{res}[R,x\langle t'\rangle],X,D=v](Hx(c)^b) \neq 0 \end{cases}$$

and hence in view (6) we see that

(18) $$\begin{cases} H_0 + Hx(c)^b \in \mathrm{Iso}(R,X\langle t'\rangle,D\geq v) \quad \text{and} \\ \mathrm{Info}[\mathrm{res}[R,x\langle t'\rangle,X,D=v](H_0+Hx(c)^b) \neq 0. \end{cases}$$

Now $\mathrm{card}(t') < \mathrm{card}(\hat{t})$ and hence, in view of (3) and (18), by the induction hypothesis we conclude that

(19) $$\mathrm{sub}[R,X=x](H_0 + Hx(c)^b) \neq 0 .$$

By (2), (15) and (19) we get

(20) $$\mathrm{sub}[R,X=x](F_1) \neq \mathrm{sub}[R,X=x](F_2).$$

Next consider the case when

(21) $$\mathrm{sub}[R,X=x](H_2) \notin \mathrm{iso}(R,x\langle t'\rangle,D\geq v).$$

Now by Lemma 3 there exists

(22) $$w \in \mathrm{inpo}(Z,D\langle t'\rangle)$$

and

(23) $$\tilde{H} \in \mathrm{Iso}(R, X\langle t'\rangle, D \geq w)$$

such that

(24) $$\mathrm{sub}[R, X = x](\tilde{H}) = \mathrm{sub}[R, X = x](H_2)$$

and

(25) $$\mathrm{Info}[\mathrm{res}[R, x\langle t'\rangle], X, D = w](\tilde{H}) \neq 0.$$

By (21), (23) and (24) we get

(26) $$w < v .$$

By (1), (23) and (25) we see that

(27) $$\begin{cases} \tilde{H}x(c)^{b+1} \in \mathrm{Iso}(R, X\langle t'\rangle, D \geq w) \quad \text{and} \\ \mathrm{Info}[\mathrm{res}[R, x\langle t'\rangle], X, D = w](\tilde{H}x(c)^{b+1}) \neq 0 . \end{cases}$$

By (6), (7), (23) and (26) we see that

$$H_0 + H_1 x(c)^b + \tilde{H}x(c)^{b+1} \in \mathrm{Iso}(R, X\langle t'\rangle, D \geq w)$$

and

$$\begin{cases} \mathrm{Info}[\mathrm{res}[R, x\langle t'\rangle], X, D = w](H_0 + H_1 x(c)^b + \tilde{H}x(c)^{b+1}) \\ = \mathrm{Info}[\mathrm{res}[R, x\langle t'\rangle], X, D = w](\tilde{H}x(c)^{b+1}) \end{cases}$$

and hence in view of (27) we get

$$(28)\qquad \begin{cases} H_0 + H_1x(c)^b + \tilde{H}x(c)^{b+1} \in \mathrm{Iso}(R,X\langle t'\rangle,D \geq w) \quad \text{and} \\ \mathrm{Info}[\mathrm{res}[R,x\langle t'\rangle],X,D=w](H_0 + H_1x(c)^b + \tilde{H}x(c)^{b+1}) \neq 0. \end{cases}$$

Now $\mathrm{card}(t') < \mathrm{card}(\hat{t})$ and hence, in view of (22) and (28), by the induction hypothesis we conclude that

$$(29)\qquad \mathrm{sub}[R,X=x](H_0 + H_1x(c)^b + \tilde{H}x(c)^{b+1}) \neq 0.$$

By (5) and (24) we have

$$\mathrm{sub}[R,X=x](F) = \mathrm{sub}[R,X=x](H_0 + H_1x(c)^b + \tilde{H}x(c)^{b+1})$$

and hence in view of (2) and (29) we see that

$$(30)\qquad \mathrm{sub}[R,X=x](F_1) \neq \mathrm{sub}[R,X=x](F_2).$$

LEMMA 5. If $x\langle t\rangle$ is R-regular then $x\langle t\rangle$ is R-quasiregular.

PROOF. Follows from Lemma 4 by taking D to be the Q-string such that $o(D) = o(x)$ and $D(c) = 1$ for all $c \in [1,o(x)]$.

§30. Initial forms for regular strings and nets

Let R be a ring. Let G' be a nonnegative ordered additive abelian semigroup and let Q' be the rational completion of G'. Let y be given where

$$\begin{cases} \text{either } y \text{ is an } R\text{-string} \\ \text{or } y \text{ is an } R\text{-net} \end{cases}$$

and let Y be given where

$$\begin{cases} Y \text{ is an indeterminate string over } R \text{ with } o(Y) = o(y) \\ \qquad\qquad \text{in case } y \text{ is an } R\text{-string} \\ \text{whereas } Y \text{ is an indeterminate net over } R \text{ with } \ell(Y) = \ell(y) \\ \qquad\qquad \text{in case } y \text{ is an } R\text{-net.} \end{cases}$$

Let E be given where

$$\begin{cases} E \in G'(o(Y)) \text{ in case } y \text{ is an } R\text{-string} \\ \text{whereas } E \in G'(\ell(Y)) \text{ in case } y \text{ is an } R\text{-net.} \end{cases}$$

Let t be given where

$$\begin{cases} t \text{ is a string-restriction in case } y \text{ is an } R\text{-string} \\ \text{whereas } t \text{ is a net-restriction in case } y \text{ is an } R\text{-net} \end{cases}$$

and let

$$\hat{t} = \begin{cases} \operatorname{supt}(o(y),t) \cap \operatorname{supt}(E) \text{ in case } y \text{ is an } R\text{-string} \\ \operatorname{supt}(\ell(y),t) \cap \operatorname{supt}(E) \text{ in case } y \text{ is an } R\text{-net.} \end{cases}$$

LEMMA 1. For any $w \in Q'$ and $P \in \{=, \geq, >, <, \leq\}$ we have

$$\mathrm{iso}(R, y\langle t\rangle, EPw) = \mathrm{iso}(R, y\langle \hat{t}\rangle, EPw).$$

PROOF. Obvious

LEMMA 2. Let

$$t' = \begin{cases} \{c' \in \hat{t} : \text{for every } c \in \hat{t} \text{ we have } n(c,c')E(c') \geq E(c) \\ \qquad \text{for some } n(c,c') \in Z\} \text{ in case } y \text{ is an R-string} \\ \{(b',c') \in \hat{t} : \text{for every } (b,c) \in \hat{t} \text{ we have} \\ \qquad n(b,c,b',c')E(b',c') \geq E(b,c) \text{ for some } n(b,c,b',c') \in Z\} \\ \qquad \text{in case } y \text{ is an R-net.} \end{cases}$$

Then

$$\bigcap_{n=1}^{\infty} y\langle t'\rangle_R^n = \bigcap_{w \in \mathrm{inpo}(Z, E\langle t'\rangle)} \mathrm{iso}(R, y\langle t'\rangle, E > w)$$

$$= \bigcap_{w \in \mathrm{inpo}(Z, E\langle t\rangle)} \mathrm{iso}(R, y\langle t\rangle, E > w) .$$

Moreover, if $\hat{t} \neq \emptyset$ then: $t' \neq \emptyset$ and

$$\bigcap_{n=1}^{\infty} y\langle t'\rangle_R^n = \bigcap_{w \in \mathrm{inpo}(Z, E\langle t'\rangle)} \mathrm{iso}(R, y\langle t'\rangle, E \geq w)$$

$$= \bigcap_{w \in \mathrm{inpo}(Z, E\langle t\rangle)} \mathrm{iso}(R, y\langle t\rangle, E \geq w).$$

PROOF. Follows from Lemma 2 of §29.

LEMMA 3. If $y\langle t\rangle$ is (R,E)-separated then given any

$$f \in R \setminus \bigcap_{w \in \mathrm{inpo}(Z,E\langle t\rangle)} \mathrm{iso}(R,y\langle t\rangle,E > w)$$

there exists $u \in \mathrm{inpo}(Z,E\langle\hat{t}\rangle)$ and $F \in \mathrm{Iso}(R,Y\langle\hat{t}\rangle,E \geq u)$ such that $\mathrm{sub}[R,Y=y](F) = f$ and $\mathrm{Info}[\mathrm{res}[R,y\langle\hat{t}\rangle],Y,E=u](F) \neq 0$.

PROOF. Follows from Lemma 3 of §29.

LEMMA 4. If $y\langle t\rangle$ is (R,E)-regular and if $u \in Q'$ and $F_1 \in \mathrm{Iso}(R,Y\langle\hat{t}\rangle,E \geq u)$ and $F_2 \in \mathrm{Iso}(R,Y\langle\hat{t}\rangle,E \geq u)$ are such that $\mathrm{sub}[R,Y=y](F_1) = \mathrm{sub}[R,Y=y](F_2)$ then we have $\mathrm{Info}[\mathrm{res}[R,y\langle\hat{t}\rangle],Y,E=u](F_1) = \mathrm{Info}[\mathrm{res}[R,y\langle\hat{t}\rangle],Y,E=u](F_2)$.

PROOF. Follows from Lemma 4 of §29.

LEMMA 5. If $y\langle t\rangle$ is (R,E)-regular then for any $u \in Q'$ we have

$$\ker(\mathrm{Info}[\mathrm{res}[R,y\langle\hat{t}\rangle],Y\langle\hat{t}\rangle,E=u]^*) \supset \ker(\mathrm{sub}[R,Y\langle\hat{t}\rangle=y,E \geq u]^*).$$

PROOF. This is simply a reformulation of Lemma 4.

DEFINITION 1. Assume that $y\langle t\rangle$ is (R,E)-regular. Then for any $u \in Q'$, in view of Lemmas 1 and 5, we can now define

$$\mathrm{info}[R,y\langle t\rangle=Y,E=u]^*: \mathrm{iso}(R,y\langle t\rangle,E \geq u) \to \mathrm{Iso}(\mathrm{res}(R,y\langle\hat{t}\rangle),Y\langle\hat{t}\rangle,E=u)$$

to be the unique $\mathrm{res}[R,y\langle\hat{t}\rangle]$-epimorphism which makes the following triangle

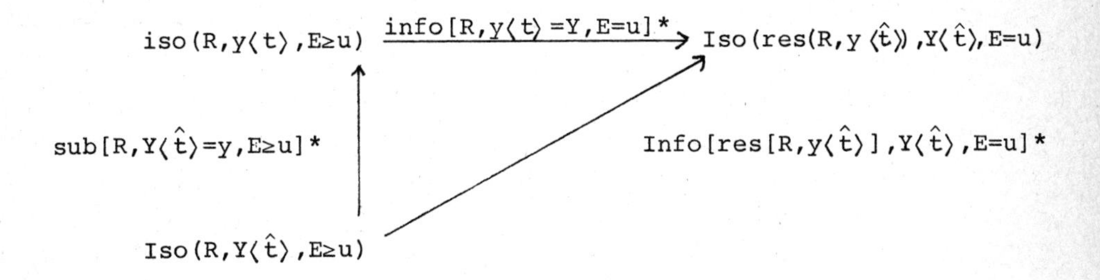

commutative, and we define

$$\text{info}[R,y\langle t\rangle=Y,E=u]: \text{iso}(R,y\langle t\rangle,E\geq u)\rightarrow \text{res}(R,y\langle \hat{t}\rangle)[Y]$$

to be the $\text{res}[R,y\langle \hat{t}\rangle]$-homomorphism induced by $\text{info}[R,y\langle t\rangle=Y,E=u]^*$, and we note that

$$\text{info}[R,y\langle t\rangle=Y,E=u] = \text{info}[R,y\langle \hat{t}\rangle=Y,E=u]$$

and

$$\text{info}[R,y\langle t\rangle=Y,E=u]^* = \text{info}[R,y\langle t\rangle=Y,E=u]^*.$$

We observe that given any positive integer n and given any $u_i \in Q'$ and $f_i \in \text{iso}(R,y\langle t\rangle,E\geq u_i)$ for $1\leq i\leq n$ we have

$$\prod_{i=1}^{n} f_i \in \text{iso}(R,y\langle t\rangle,E\geq \sum_{i=1}^{n} u_i)$$

and

$$\text{info}[R,y\langle t\rangle=Y,E=\sum_{i=1}^{n} u_i]\left(\prod_{i=1}^{n} f_i\right) = \prod_{i=1}^{n} \text{info}[R,y\langle t\rangle,E=u_i](f_i).$$

For any

$$f \in R \setminus \bigcap_{w \in \mathrm{inpo}(Z,E\langle t\rangle)} \mathrm{iso}(R,y\langle t\rangle,E > w)$$

in view of Lemmas 1, 3 and 4 we can now define

$$\begin{aligned} \mathrm{ord}[R,y\langle t\rangle,E](f) = {} & \text{the unique } u' \in Q' \text{ such that} \\ & f \in \mathrm{iso}(R,y\langle t\rangle,E \geq u') \text{ and} \\ & \mathrm{info}(R,y\langle t\rangle,E = u')(f) \neq 0 \end{aligned}$$

and we observe that then

$$\mathrm{ord}[R,y\langle t\rangle,E](f) \in \mathrm{inpo}(Z,E\langle \hat{t}\rangle)$$

and for any $u \in Q'$ we have:

$$u \leq \mathrm{ord}[R,y\langle t\rangle,E](f) \Leftrightarrow f \in \mathrm{iso}(R,y\langle t\rangle,E \geq u)$$

whereas:

$$u < \mathrm{ord}[R,y\langle t\rangle,E](f) \Leftrightarrow \begin{cases} f \in \mathrm{iso}(R,y\langle t\rangle,E \geq u) \text{ and} \\ \mathrm{info}[R,y\langle t\rangle = Y,E = u](f) = 0. \end{cases}$$

We also put

$$\mathrm{ord}[R,y\langle t\rangle,E](f) = \infty \text{ for all } f \in \bigcap_{w \in \mathrm{inpo}(Z,E\langle t\rangle)} \mathrm{iso}(R,y\langle t\rangle,E > w).$$

We note that for any $f \in R$ we now have:

$$\mathrm{ord}[R,y\langle t\rangle,E](f) = \infty \Leftrightarrow f \in \bigcap_{w \in \mathrm{inpo}(Z,E\langle t\rangle)} \mathrm{iso}(R,y\langle t\rangle,E > w).$$

We also observe that for any $f \in R$ we have

$$\operatorname{ord}[R,y\langle t\rangle,E](f) = \operatorname{ord}[R,y\langle\hat{t}\rangle,E](f).$$

For any $f' \subset R$ we put

$$\operatorname{ord}[R,y\langle t\rangle,E](f') = \{\operatorname{ord}[R,y\langle t\rangle,E](f): f \in f'\}.$$

We define the map

$$\operatorname{info}[R,y\langle t\rangle=Y,E]: R \to \operatorname{res}(R,y\langle\hat{t}\rangle)[Y]$$

by putting, for all $f \in R$,

$$\operatorname{info}[R,y\langle t\rangle = Y,E](f)$$

$$= \begin{cases} \operatorname{info}[R,y\langle t\rangle=Y,E=\operatorname{ord}[R,y\langle t\rangle,E](f)](f) \text{ if } \operatorname{ord}[R,y\langle t\rangle,E](f) \neq \infty \\ 0 \quad \text{if} \quad \operatorname{ord}[R,y\langle t\rangle,E](f) = \infty. \end{cases}$$

We define

$$\operatorname{info}[R,y\langle t\rangle = Y,E]^*: R \to \operatorname{res}(R,y\langle\hat{t}\rangle)[Y\langle\hat{t}\rangle]$$

to be the surjective map induced by $\operatorname{info}[R,y\langle t\rangle=Y,E]$. We observe that

$$\operatorname{info}[R,y\langle t\rangle=Y,E]^* = \operatorname{info}[R,y\langle\hat{t}\rangle=Y,E]^*$$

and

$$\operatorname{info}[R,y\langle t\rangle=Y,E] = \operatorname{info}[R,y\langle\hat{t}\rangle=Y,E].$$

For any $f' \subset R$ we define

$$\text{info}[R,y\langle t\rangle=Y,E]^*((f')) = \text{the ideal in } \text{res}(R,y\langle\hat{t}\rangle)[Y\langle\hat{t}\rangle]$$
$$\text{generated by } \text{info}[R,y\langle t\rangle=Y,E]^*(f')$$

and

$$\text{info}[R,y\langle t\rangle=Y,E]((f')) = \text{the ideal in } \text{res}(R,y\langle\hat{t}\rangle)[Y]$$
$$\text{generated by } \text{info}[R,y\langle t\rangle=Y,E](f').$$

Given any I where

$$\begin{cases} \text{either } I \text{ is an ideal in } R \text{ with } y\langle\hat{t}\rangle_R^1 \subset I \\ \text{or } I = \bar{x} \text{ where } \bar{x} \text{ is an } R\text{-string with } y\langle\hat{t}\rangle_R^1 \subset \bar{x}_R^1 \\ \text{or } I = \bar{x}\langle\bar{t}\rangle \text{ where } \bar{x} \text{ is an } R\text{-string and } \bar{t} \text{ is a string-} \\ \qquad \text{restriction with } y\langle\hat{t}\rangle_R^1 \subset \bar{x}\langle\bar{t}\rangle_R^1 \\ \text{or } I = \bar{y} \text{ is an } R\text{-net with } y\langle\hat{t}\rangle_R^1 \subset \bar{y}_R^1 \\ \text{or } I = \bar{y}\langle\bar{t}\rangle \text{ where } \bar{y} \text{ is an } R\text{-net and } \bar{t} \text{ is a net-restriction} \\ \qquad \text{with } y\langle\hat{t}\rangle_R^1 \subset \bar{y}\langle\bar{t}\rangle_R^1 \end{cases}$$

we define

$$\text{info}[(R,I),y\langle t\rangle=Y,E]: R \to \text{res}(R,I)[Y]$$

to be the composition of the maps

$$\begin{array}{ccc} R \xrightarrow{\text{info}[R,y\langle t\rangle=Y,E]} & \text{res}(R,y\langle\hat{t}\rangle)[Y] & \\ & \Big\downarrow & \text{Sub}[\text{res}[(R,I),y\langle\hat{t}\rangle],Y] \\ & \text{res}(R,I)[Y] & \end{array}$$

and we define

$$\text{info}[(R,I),y\langle t\rangle=Y,E]^*:\ R \to \text{res}(R,I)[Y\langle\hat{t}\rangle]$$

to be the surjective map induced by info[(R,I),y⟨t⟩=Y,E] and we observe that

$$\text{info}[(R,I),y\langle t\rangle=Y,E]^* = \text{info}[(R,I),y\langle\hat{t}\rangle=Y,E]^*$$

and

$$\text{info}[(R,I),y\langle t\rangle=Y,E]=\text{info}[(R,I),y\langle\hat{t}\rangle=Y,E]$$

and for any $f' \subset R$ we define

$$\text{info}[(R,I),y\langle t\rangle=Y,E]^*((f')) = \text{the ideal in res}(R,I)[Y\langle\hat{t}\rangle] \text{ generated by info}[(R,I),y\langle t\rangle=Y,E]^*(f')$$

and

$$\text{info}[(R,I),y\langle t\rangle=Y,E]((f')) = \text{the ideal in res}(R,I)[Y] \text{ generated by info}[(R,I),y\langle t\rangle=Y,E](f')$$

and for any $u \in Q'$ we define the res[R,I]-homomorphism

$$\text{info}[(R,I),y\langle t\rangle=Y,E=u]:\ \text{iso}(R,y\langle t\rangle,E\geq u) \to \text{res}(R,I)[Y]$$

to be the composition of the maps

$$\begin{array}{ccc}\text{iso}(R,y\langle t\rangle,E\geq u) & \xrightarrow{\text{info}[R,y\langle t\rangle=Y,E=u]} & \text{res}(R,y\langle\hat{t}\rangle)[Y] \\ & & \Big\downarrow \text{Sub}[\text{res}[(R,I),y\langle\hat{t}\rangle],Y] \\ & & \text{res}(R,I)[Y]\end{array}$$

and we define

$$\mathrm{info}[(R,I),y\langle t\rangle=Y,E=u]^*: \mathrm{iso}(R,y\langle t\rangle,E\geq u) \to \mathrm{Iso}(\mathrm{res}(R,I),Y\langle\hat{t}\rangle,E=u)$$

to be the $\mathrm{res}[R,I]$-epimorphism induced by $\mathrm{info}[(R,I),y\langle t\rangle=Y,E=u]$, and we observe that

$$\mathrm{info}[(R,I),y\langle t\rangle=Y,E=u] = \mathrm{info}[(R,I),y\langle\hat{t}\rangle=Y,E=u]$$

and

$$\mathrm{info}[(R,I),y\langle t\rangle=Y,E=u]^* = \mathrm{info}[(R,I),y\langle\hat{t}\rangle=Y,E=u]^*.$$

If $y\langle\hat{t}\rangle_R^1 \subset M(R)$ then we define the map

$$\mathrm{info}[(R),y\langle t\rangle=Y,E]: R \to \mathrm{res}(R)[Y]$$

by putting

$$\mathrm{info}[(R),y\langle t\rangle=Y,E] = \mathrm{info}[(R,M(R)),y\langle t\rangle=Y,E]$$

and we define

$$\mathrm{info}[(R),y\langle t\rangle=Y,E]^*: R \to \mathrm{res}(R)[Y\langle\hat{t}\rangle]$$

to be the surjective map induced by $\mathrm{info}[(R),y\langle t\rangle=Y,E]$ and we observe that

$$\mathrm{info}[(R),y\langle t\rangle=Y,E]^* = \mathrm{info}[(R),y\langle\hat{t}\rangle=Y,E]^*$$

and

$$\mathrm{info}[(R),y\langle t\rangle=Y,E] = \mathrm{info}[(R),y\langle\hat{t}\rangle=Y,E]$$

and for any $f' \subset R$ we define

$$\text{info}[(R),y\langle t\rangle=Y,E]^*((f')) = \text{the ideal in } \text{res}(R)[Y\langle\hat{t}\rangle] \text{ generated by } \text{info}[(R),y\langle t\rangle=Y,E]^*(f')$$

and

$$\text{info}[(R),y\langle t\rangle=Y,E]((f')) = \text{the ideal in } \text{res}(R)[Y] \text{ generated by } \text{info}[(R),y\langle t\rangle=Y,E](f')$$

and for any $u \in Q'$ we define the res[R]-homomorphism

$$\text{info}[(R),y\langle t\rangle=Y,E=u]: \text{iso}(R,y\langle t\rangle,E\geq u) \to \text{res}(R)[Y]$$

by putting

$$\text{info}[(R),y\langle t\rangle=Y,E=u] = \text{info}[(R,M(R)),y\langle t\rangle = Y,E = u]$$

and we define

$$\text{info}[(R),y\langle t\rangle=Y,E=u]^*: \text{iso}(R,y\langle t\rangle,E\geq u) \to \text{Iso}(\text{res}(R),Y\langle\hat{t}\rangle,E=u)$$

to be the res[R]-epimorphism induced by $\text{info}[(R),y\langle t\rangle=Y,E=u]$, and we observe that

$$\text{info}[(R),y\langle t\rangle=Y,E=u] = \text{info}[(R),y\langle\hat{t}\rangle=Y,E=u]$$

and

$$\text{info}[(R),y\langle t\rangle=Y,E=u]^* = \text{info}[(R),y\langle\hat{t}\rangle=Y,E=u]^*.$$

DEFINITION 2. If y is (R,E)-regular then we verbatim take over the entire above material of Definition 1 after everywhere deleting $\langle t\rangle$ and replacing $\hat{t}$ by E (or, equivalently, by supt(E)).

LEMMA 6. If $y\langle t\rangle$ is R-regular then $y\langle t\rangle$ is R-quasiregular.

PROOF. Follows from Lemma 5 of §29.

LEMMA 7. If $y\langle t\rangle$ is R-separated and R-quasiregular then $y\langle t\rangle$ is R-regular.

PROOF. This is only a repetition of Lemma 3 of §25.

LEMMA 8. $y\langle t\rangle$ is R-superregular

$\Leftrightarrow$ $y\langle t\rangle$ is R-ultraseparated and R-regular

$\Leftrightarrow$ $y\langle t\rangle$ is R-ultraseparated and R-ultraregular

$\Leftrightarrow$ $y\langle t\rangle$ is R-ultraseparated and R-ultraquasiregular

$\Leftrightarrow$ $y\langle t\rangle$ is R-ultraseparated and R-quasiregular.

PROOF. Follows from Lemmas 6 and 7.

§31. Protochips and parachips

Recall that for any prechip e we have put

$$e[B] = e(B,B,0) \quad \text{for} \quad 1 \le B \le o(\ell(e))$$

and

$$e[[B]] = \sum_{\substack{1\le b\le o(\ell(e)) \\ 0\le c\le b(\ell(e))}} e(B,b,c) \quad \text{for} \quad 1 \le B \le o(\ell(e)).$$

By a *protochip* we mean a prechip e such that

$$e[[B]] \ne 0 \quad \text{for} \quad 1 \le B \le o(\ell(e)).$$

By a *parachip* we mean a protochip e such that

$$e[B] \ne 0 \quad \text{for} \quad 1 \le B \le o(\ell(e)) - 2 .$$

§32. N-support of an indexing string for $2 \le N \le 6$

Let ℓ be an indexing string. Recall that

$$\mathrm{supt}(\ell) = \{(b,c) \in Z^{(2)}: 1 \le b \le o(\ell) \text{ and } 1 \le c \le b(\ell)\}.$$

We define

$$\mathrm{supt}_2(\ell) = \{(b,c) \in Z^{(2)}: 1 \le b \le o(\ell) \text{ and } 0 \le c \le b(\ell)\}$$

and

$$\mathrm{supt}_3(\ell) = \{(B,b,c) \in Z^{(3)}: 1 \le B \le o(\ell) \text{ and } (b,c) \in \mathrm{supt}_2(\ell)\}$$

and

$$\mathrm{supt}_4(\ell) = \{(B,\tilde{B},\hat{B},\hat{C}) \in Z^{(4)}: B \in [1,o(\ell)] \text{ and } (\tilde{B},\hat{B},\hat{C}) \in \mathrm{supt}_3(\ell)\}$$

and

$$\mathrm{supt}_5(\ell) = \{(\tilde{B},\hat{B},\hat{C},b,c) \in Z^{(5)}: (\tilde{B},\hat{B},\hat{C}) \in \mathrm{supt}_3(\ell) \text{ and } (b,c) \in \mathrm{supt}_2(\ell)\}$$

and

$$\mathrm{supt}_6(\ell) = \{(B,\tilde{B},\hat{B},\hat{C},b,c) \in Z^{(6)}: B \in [1,o(\ell)] \text{ and } (\tilde{B},\hat{B},\hat{C},b,c) \in \mathrm{supt}_5(\ell)\}.$$

§33. Prescales

By a <u>prescale</u> we mean a system E consisting of: an indexing string $\ell(E)$ called the index of E, and for every $(B,\tilde{B},\hat{B},\hat{C},b,c) \in \text{supt}_6(\ell(E))$

a nonnegative rational number $E(B,\tilde{B},\hat{B},\hat{C},b,c)$

called the $(B,\tilde{B},\hat{B},\hat{C},b,c)^{\text{th}}$ <u>primary component</u> of E, and for every $(B,\tilde{B},\hat{B},\hat{C},b,c) \in \text{supt}_6(\ell(E))$

a nonnegative rational number $E((B,\tilde{B},\hat{B},\hat{C},b,c))$

called the $(B,\tilde{B},\hat{B},\hat{C},b,c)^{\text{th}}$ <u>secondary component</u> of E.

We define

$$\text{denom}(E) = \{0 \neq n \in Z\colon nE(B,\tilde{B},\hat{B},\hat{C},b,c)) \in Z \text{ for all } (B,\tilde{B},\hat{B},\hat{C},b,c) \in \text{supt}_6(\ell(E))\}$$

and

$$\text{denom}((E)) = \{0 \neq n \in Z\colon nE((B,\tilde{B},\hat{B},\hat{C},b,c)) \in Z \text{ for all } (B,\tilde{B},\hat{B},\hat{C},b,c) \in \text{supt}_6(\ell(E))\}$$

and for any set E' of prescales we put

$$\text{denom}(E') = \bigcap_{E \in E'} \text{denom}(E)$$

and

$$\mathrm{denom}((E')) = \bigcap_{E \in E'} \mathrm{denom}((E)).$$

For any $B \in [1, o(\ell(E))]$ we define

$$\mathrm{denom}(E,B) = \{0 \neq n \in Z\colon nE(B,\tilde{B},\hat{B},\hat{C},b,c) \in Z \text{ for all } (\tilde{B},\hat{B},\hat{C},b,c) \in \mathrm{supt}_5(\ell(E))\}$$

and

$$\mathrm{denom}((E,B)) = \{0 \neq n \in Z\colon nE((B,\tilde{B},\hat{B},\hat{C},b,c)) \in Z \text{ for all } (\tilde{B},\hat{B},\hat{C},b,c) \in \mathrm{supt}_5(\ell(E))\}.$$

For any $u \in Q$, by uE we denote the prescale with $\ell(uE) = \ell(E)$ such that $(uE)(B,\tilde{B},\hat{B},\hat{C},b,c) = uE(B,\tilde{B},\hat{B},\hat{C},b,c)$ and $(uE)((B,\tilde{B},\hat{B},\hat{C},b,c) = uE((B,\tilde{B},\hat{B},\hat{C},b,c))$ for all $(B,\tilde{B},\hat{B},\hat{C},b,c) \in \mathrm{supt}_6(\ell(E))$. Likewise, given any indexing string ℓ, we may regard the set of all prescales whose index is ℓ as an additive abelian semigroup with componentwise addition.

Finally, given any $(B,\tilde{B},\hat{B},\hat{C}) \in \mathrm{supt}_4(\ell(E))$, by $E(B,\tilde{B},\hat{B},\hat{C})$ we denote the Q-net whose index is $\ell(E)$ and whose $(b,c)^{\text{th}}$ component is $E(B,\tilde{B},\hat{B},\hat{C},b,c)$ for all $(b,c) \in \mathrm{supt}(\ell(E))$, and by $E((B,\tilde{B},\hat{B},\hat{C}))$ we denote the Q-net whose index is $\ell(E)$ and whose $(b,c)^{\text{th}}$ component is $E((B,\tilde{B},\hat{B},\hat{C},b,c))$ for all $(b,c) \in \mathrm{supt}(\ell(E))$.

§34. Derived prescales

Given any prechip e, by e^* we denote the prescale for which $\ell(e^*) = \ell(e)$ and whose components are defined as follows. Firstly we put

$$e^*(B,\tilde{B},\hat{B},\hat{C},b,c)$$

$$= e^*((\tilde{B},\hat{B},\hat{C},b,c)) = 0 \text{ if } (B,\tilde{B},\hat{B},\hat{C},b,c) \in \text{supt}_6(\ell(e)) \text{ and } b < B.$$

Secondly we put

$$e^*(B,\tilde{B},\hat{B},\hat{C},b,c) = \begin{cases} 0 \text{ if } (B,\tilde{B},\hat{B},\hat{C},b,c) \in \text{supt}_6(\ell(e)) \text{ and } b \geq B > \tilde{B} \\ 0 \text{ if } (B,\tilde{B},\hat{B},\hat{C},b,c) \in \text{supt}_6(\ell(e)) \text{ and } b \geq B \leq \tilde{B} > \hat{B}. \end{cases}$$

Thirdly, given any

$$(B,\tilde{B},\hat{B},\hat{C}) \in \text{supt}_4(\ell(e)) \quad \text{with} \quad B \leq \tilde{B} \leq \hat{B}$$

to begin with we put

$$e^*(B,\tilde{B},\hat{B},\hat{C},b,c) = \begin{cases} 0 & \text{if } \hat{B} < b \leq o(\ell(e)) \text{ and } 0 \leq c \leq b(\ell(e)) \\ 0 & \text{if } \tilde{B} \leq b < \hat{B} \text{ and } 0 \leq c \leq b(\ell(e)) \\ 0 & \text{if } b = \hat{B} \text{ and } c \neq \hat{C} \neq 0 \leq c \leq b(\ell(e)) \\ 1 & \text{if } b = \hat{B} \text{ and } c = \hat{C} \neq 0 \\ 0 & \text{if } \tilde{B} < b = \hat{B} \text{ and } \hat{C} = 0 \leq c \leq b(\ell(e)) \\ 0 & \text{if } \tilde{B} = b = \hat{B} \leq o(\ell(e)) - 2 \text{ and } \hat{C} = 0 \leq c \leq b(\ell(e)) \\ 1 & \text{if } \tilde{B} = b = \hat{B} \geq o(\ell(e)) - 1 \text{ and } \hat{C} = 0 \leq c \leq b(\ell(e)) \end{cases}$$

and then, by decreasing induction on b, we define

$$e^*(B,\tilde{B},\hat{B},\hat{C},b,c) = \sum_{\substack{b+1\le b'\le o(\ell(e)) \\ 0\le c'\le b'(\ell(e))}} e^*(B,\tilde{B},\hat{B},\hat{C},b',c')e(b+1,b',c')$$

$$\text{if } \begin{cases} B \le b < \tilde{B} \text{ and} \\ 0 \le c \le b(\ell(e)). \end{cases}$$

Finally we define

$$e^*((B,\tilde{B},\hat{B},\hat{C},b,c)) = \sum_{\substack{b\le b'\le o(\ell(e)) \\ 0\le c'\le b'(\ell(e))}} e^*(B,\tilde{B},\hat{B},\hat{C},b',c')e(b,b',c')$$

$$\text{if } \begin{cases} (B,\tilde{B},\hat{B},\hat{C},b,c) \in \text{supt}_6(\ (e)) \\ \text{and } B \le b. \end{cases}$$

§35. Supports of prescales

Given any prescale E we define

$$\mathrm{supt}(E) = \{(B,\widetilde{B},\hat{B},\hat{C}) \in \mathrm{supt}_4(\ell(E)): E(B,\widetilde{B},\hat{B},\hat{C},b,c) \neq 0 \text{ for some } (b,c) \in \mathrm{supt}_2(\ell(E))\}$$

and

$$\mathrm{supt}((E)) = \{(B,\widetilde{B},\hat{B},\hat{C}) \in \mathrm{supt}_4(\ell(E)):\} \quad E((B,\widetilde{B},\hat{B},\hat{C},b,c)) \neq 0 \text{ for some } (b,c) \in \mathrm{supt}_2(\ell(E))$$

and

for every $B \in [1,o(\ell(E))]$

we define

$$\mathrm{supt}(E,B) = \{(\widetilde{B},\hat{B},\hat{C}) \in \mathrm{supt}_3(\ell(E)): (B,\widetilde{B},\hat{B},\hat{C}) \in \mathrm{supt}(E)\}$$

and

$$\mathrm{supt}((E,B)) = \{(\widetilde{B},\hat{B},\hat{C}) \in \mathrm{supt}_3(\ell(E)): (B,\widetilde{B},\hat{B},\hat{C}) \in \mathrm{supt}((E))\}.$$

§36. Protoscales

By a <u>protoscale</u> we mean a prescale E such that

$$E(B,\widetilde{B},\hat{B},\hat{C},B,0) = 1 \quad \text{for all} \quad (B,\widetilde{B},\hat{B},\hat{C}) \in \text{supt}(E).$$

Given any prescale E, we define E^* to be the prescale such that $\ell(E^*) = \ell(E)$ and such that for every $(B,\widetilde{B},\hat{B},\hat{C},b,c) \in \text{supt}_6(\ell(E))$ we have

$$E^*(B,\widetilde{B},\hat{B},\hat{C},b,c) = \begin{cases} \dfrac{E(B,\widetilde{B},\hat{B},\hat{C},b,c)}{E(B,\widetilde{B},\hat{B},\hat{C},B,0)} & \text{if } E(B,\widetilde{B},\hat{B},\hat{C},B,0) \neq 0 \\ 0 \quad \text{if } E(B,\widetilde{B},\hat{B},\hat{C},B,0) = 0 \end{cases}$$

and

$$E^*((B,\widetilde{B},\hat{B},\hat{C},b,c)) = \begin{cases} \dfrac{E(B,\widetilde{B},\hat{B},\hat{C},b,c)}{E((B,\widetilde{B},\hat{B},\hat{C},B,0))} & \text{if } E((B,\widetilde{B},\hat{B},\hat{C},B,0)) \neq 0 \\ 0 \quad \text{if } E((B,\widetilde{B},\hat{B},\hat{C},B,0)) = 0 \end{cases} .$$

We note that if E is any prescale then obviously E^* is a protoscale.

It follows that if e is any prechip then e^{**} is a protoscale and we have $\ell(e^{**}) = \ell(e)$ and for every $(B,\widetilde{B},\hat{B},\hat{C},b,c) \in \text{supt}_6(\ell(e))$ we have

$$e^{**}(B,\widetilde{B},\hat{B},\hat{C},b,c) = \begin{cases} \dfrac{e^*(B,\widetilde{B},\hat{B},\hat{C},b,c)}{e^*(B,\widetilde{B},\hat{B},\hat{C},B,0)} & \text{if } e^*(B,\widetilde{B},\hat{B},\hat{C},B,0) \neq 0 \\ 0 \quad \text{if } e^*(B,\widetilde{B},\hat{B},\hat{C},B,0) = 0 \end{cases}$$

and

$$e^{**}((B,\widetilde{B},\hat{B},\hat{C},b,c)) = \begin{cases} \dfrac{e^{*}(B,\widetilde{B},\hat{B},\hat{C},b,c)}{e^{*}((B,\widetilde{B},\hat{B},\hat{C},B,0))} & \text{if } e^{*}((B,\widetilde{B},\hat{B},\hat{C},B,0)) \neq 0 \\ 0 \quad \text{if} \quad e^{*}((B,\widetilde{B},\hat{B},\hat{C},B,0)) = 0 \ . \end{cases}$$

§37. Inner products for protoscales

Given any protoscale E and given any $B \in [1, o(\ell(E))]$ and $G \subset Q$ and $u \in Q$ we define

$$G(E(B) \geq u) = \Big\{ j \in G(\ell(E)) : \ \operatorname{inpo}(j, E(B, \tilde{B}, \hat{B}, \hat{C})) \geq u \ \text{ for all } (\tilde{B}, \hat{B}, \hat{C}) \in \operatorname{supt}(E, B) \Big\}$$

and

$$G(E((B)) \geq u) = \Big\{ j \in G(\ell(E)) : \operatorname{inpo}(j, E((B, \tilde{B}, \hat{B}, \hat{C}))) \geq u \ \text{ for all } (\tilde{B}, \hat{B}, \hat{C}) \in \operatorname{supt}((E, B)) \Big\}$$

and for any $P \in \{>, =\}$ we define

$$G(E(B)Pu) = \Big\{ j \in G(E(B) \geq u) : \operatorname{inpo}(j, E(B, o(\ell(E)), o(\ell(E)), 0))Pu \Big\}$$

and

$$G(E((B))Pu) = \Big\{ j \in G(E((B)) \geq u) : \operatorname{inpo}(j, E((B, o(\ell(E)), o(\ell(E)), 0)))Pu \Big\} .$$

§38. Scales and isobars

By a <u>scale</u> we mean a protoscale E such that

$$(1)\quad \begin{cases} E((B,\widetilde{B},\hat{B},\hat{C},b,c)) = 0 = E(B,\widetilde{B},\hat{B},\hat{C},b,c) \\ \text{for all those } (B,\widetilde{B},\hat{B},\hat{C},b,c) \in \operatorname{supt}_6(\ell(E)) \text{ for which } b < B \end{cases}$$

and

$$(2)\quad \begin{cases} E((B,o(\ell(E)),o(\ell(E)),0,b,c)) \neq 0 \neq E(B,o(\ell(E)),o(\ell(E)),0,b,c) \\ \text{for all those } (B,b,c) \in \operatorname{supt}_3(\ell(E)) \text{ for which } b \geq B. \end{cases}$$

We note that for any scale E we obviously have

$$\begin{cases} (o(\ell(E)),o(\ell(E)),0) \in \operatorname{supt}(E,B) \cap \operatorname{supt}((E,B)) \\ \text{for all } B \in [1,o(\ell(E))]. \end{cases}$$

Now let E be a scale, let R be a ring, let Y be an indeterminate net over R with $\ell(Y) = \ell(E)$, and let $u \in Q$.

For any

$$B \in [1,o(\ell(E))-1] \quad \text{and} \quad B' \in [1,B]$$

we define the

ideals $Y\langle B',B\rangle^u_{(R,E\geq)Q}$ and $Y\langle B',B\rangle^u_{((R,E\geq))Q}$ in $R[Y\langle B'\rangle]_Q$

by putting

$$Y\langle B',B\rangle^u_{(R,E\geq)Q} = \bigcap_{(\widetilde{B},\hat{B},\hat{C})\in\operatorname{supt}(E,B)} \operatorname{Iso}(R,Y\langle B'\rangle,E(B,\widetilde{B},\hat{B},\hat{C}) \geq u)_Q$$

and

$$Y\langle B',B\rangle^{u}_{((R,E\geq))Q} = \bigcap_{(\tilde{B},\hat{B},\hat{C})\in\mathrm{supt}((E,B))} \mathrm{Iso}(R,Y\langle B'\rangle,E((B,\tilde{B},\hat{B},\hat{C}))\geq u)_{Q}$$

and we define the

ideals $Y\langle B',B\rangle^{u}_{(R,E\geq)}$ and $Y\langle B',B\rangle^{u}_{((R,E\geq))}$ in $R[Y\langle B'\rangle]$

by putting

$$Y\langle B',B\rangle^{u}_{(R,E\geq)} = Y\langle B',B\rangle^{u}_{(R,E\geq)Q} \cap R[Y\langle B'\rangle]$$

and

$$Y\langle B',B\rangle^{u}_{((R,E\geq))} = Y\langle B',B\rangle^{u}_{((R,E\geq))Q} \cap R[Y\langle B'\rangle]$$

and given any $R_0 \subset R$ with $0 \in R_0$ we put

$$\mathrm{Iso}(R_0,Y\langle B',B\rangle,E\geq u)_Q = Y\langle B',B\rangle^{u}_{(R,E\geq)Q} \cap R_0[Y\langle B'\rangle]_Q$$

$$\mathrm{Iso}((R_0,Y\langle B',B\rangle,E\geq u))_Q = Y\langle B',B\rangle^{u}_{((R,E\geq))Q} \cap R_0[Y\langle B'\rangle]_Q$$

$$\mathrm{Iso}(R_0,Y\langle B',B\rangle,E\geq u) = Y\langle B',B\rangle^{u}_{(R,E\geq)} \cap R_0[Y\langle B'\rangle]$$

and

$$\mathrm{Iso}((R_0,Y\langle B',B\rangle,E\geq u)) = Y\langle B',B\rangle^{u}_{((R,E\geq))} \cap R_0[Y\langle B'\rangle].$$

We observe that, in view of (1),

$$(3)\begin{cases} \text{for any integers } B,B',B'' \text{ with } 1 \le B'' \le B' \le B \le o(\ell(e))-1 \text{ we have:} \\ Y\langle B'',B\rangle^{u}_{(R,E\geq)Q} \cap R[Y\langle B'\rangle]_Q = Y\langle B',B\rangle^{u}_{(R,E\geq)Q} \\ Y\langle B'',B\rangle^{u}_{(R,E\geq)Q} = Y\langle B',B\rangle^{u}_{(R,E\geq)Q} R[Y\langle B''\rangle]_Q \\ Y\langle B'',B\rangle^{u}_{((R,E\geq))Q} \cap R[Y\langle B'\rangle]_Q = Y\langle B',B\rangle^{u}_{((R,E\geq))Q} \\ Y\langle B'',B\rangle^{u}_{((R,E\geq))Q} = Y\langle B',B\rangle^{u}_{((R,E\geq))Q} R[Y\langle B''\rangle]_Q \end{cases}$$

and

$$(4)\begin{cases} \text{for any integers } B,B',B'' \text{ with } 1 \le B'' \le B' \le B \le o(\ell(e))-1 \text{ we have:} \\ Y\langle B'',B\rangle^{u}_{(R,E\geq)} \cap R[Y\langle B'\rangle] = Y\langle B',B\rangle^{u}_{(R,E\geq)} \\ Y\langle B'',B\rangle^{u}_{(R,E\geq)} = Y\langle B',B\rangle^{u}_{(R,E\geq)} R[Y\langle B''\rangle] \\ Y\langle B'',B\rangle^{u}_{((R,E\geq))} \cap R[Y\langle B'\rangle] = Y\langle B',B\rangle^{u}_{((R,E\geq))} \\ Y\langle B'',B\rangle^{u}_{((R,E\geq))} = Y\langle B',B\rangle^{u}_{((R,E\geq))} R[Y\langle B''\rangle]. \end{cases}$$

For any

$$B \in [1,o(\ell(e))-1] \quad \text{and} \quad B' \in [1,B] \quad \text{and} \quad P \in \{=,>\}$$

we define the

R-submodules $Y\langle B',B\rangle^{u}_{(R,EP)Q}$ and $Y\langle B',B\rangle^{u}_{((R,EP))Q}$ of $R[Y\langle B'\rangle]_Q$

by putting

$$Y\langle B',B\rangle^{u}_{(R,EP)Q}$$

$$= Y\langle B',B\rangle^{u}_{(R,E\geq)Q} \cap \mathrm{Iso}(R,Y\langle B\rangle,E(B,o(\ell(E)),o(\ell(E)),0)Pu)_{Q}$$

and

$$Y\langle B',B\rangle^{u}_{((R,EP))Q}$$

$$= Y\langle B',B\rangle^{u}_{((R,E\geq))Q} \cap \mathrm{Iso}(R,Y\langle B\rangle,E((B,o(\ell(E)),o(\ell(E)),0))Pu)_{Q}$$

and we define the

R-submodules $Y\langle B',B\rangle^{u}_{(R,EP)}$ and $Y\langle B',B\rangle^{u}_{((R,EP))}$ of $R[Y\langle B'\rangle]$

by putting

$$Y\langle B',B\rangle^{u}_{(R,EP)} = Y\langle B',B\rangle^{u}_{(R,EP)Q} \cap R[Y\langle B'\rangle]$$

and

$$Y\langle B',B\rangle^{u}_{((R,EP))} = Y\langle B',B\rangle^{u}_{((R,EP))Q} \cap R[Y\langle B'\rangle]$$

and given any $R_0 \subset R$ with $0 \in R_0$ we put

$$\mathrm{Iso}(R_0,Y\langle B',B\rangle,EPu)_{Q} = Y\langle B',B\rangle^{u}_{(R,EP)Q} \cap R_0[Y\langle B'\rangle]_{Q}$$

$$\mathrm{Iso}((R_0,Y\langle B',B\rangle,EPu))_{Q} = Y\langle B',B\rangle^{u}_{((R,EP))Q} \cap R_0[Y\langle B'\rangle]_{Q}$$

$$\mathrm{Iso}(R_0,Y\langle B',B\rangle,EPu) = Y\langle B',B\rangle^{u}_{(R,EP)} \cap R_0[Y\langle B'\rangle]$$

and

$$\text{Iso}((R_0, Y\langle B',B\rangle, \text{EPu})) = Y\langle B',B\rangle^{u}_{((R,EP))} \cap R_0[Y\langle B'\rangle].$$

We note that obviously

$$(5)\quad \begin{cases} \text{for any } B \in [1,o(\ell(e))-1] \text{ and } B' \in [1,B] \text{ and } P \in \{=,>\} \text{ we have:} \\ Y\langle B',B\rangle^{u}_{(R,EP)Q} = Y\langle B,B\rangle^{u}_{(R,EP)Q} \\ \text{and} \\ Y\langle B',B\rangle^{u}_{((R,EP))Q} = Y\langle B,B\rangle^{u}_{((R,EP))Q} \end{cases}$$

and

$$(6)\quad \begin{cases} \text{for any } B \in [1,o(\ell(e))-1] \text{ and } B' \in [1,B] \text{ and } P \in \{=,>\} \text{ we have:} \\ Y\langle B',B\rangle^{u}_{(R,EP)} = Y\langle B,B\rangle^{u}_{(R,EP)} \\ \text{and} \\ Y\langle B',B\rangle^{u}_{((R,EP))} = Y\langle B,B\rangle^{u}_{((R,EP))}. \end{cases}$$

§39. Properties of derived prescales

This section is by way of details of proofs of assertions to be made in the next section. The reader may decide how much of these details he wishes to read.

Let e be a protochip.

Recall that by the definition of a protochip

(1) $e(B,b,c) \in Q$ for every $(B,b,c) \in \text{supt}_3(\ell(e))$

(2) $e(B,b,c) = 0$ if $1 \le b \le B = o(\ell(e))$ and $1 \le c \le b(\ell(e))$

(3) $e(B,b,c) = 0$ if $1 \le b < B \le o(\ell(e))$ and $0 \le c \le b(\ell(e))$

(4) $e(B,b,0) = 0$ if $1 \le B < b \le o(\ell(e))$

(5) $e[B] = e(B,B,0)$ for $1 \le B \le o(\ell(e))$

(6) $e[[B]] = \sum_{\substack{1 \le b \le o(\ell(e)) \\ 0 \le c \le b(\ell(e))}} e(B,b,c) \neq 0$ for $1 \le B \le o(\ell(e))$

and

(7) $\begin{cases} \text{for } 1 \le B \le o(\ell(e)), \text{ by } e(B) \text{ we are denoting the } Q\text{-net} \\ \text{whose index is } \ell(e) \text{ and whose } (b,c)^{\text{th}} \text{ component is} \\ e(B,b,c) \text{ for } 1 \le b \le o(\ell(e)) \text{ and } 1 \le c \le b(\ell(e)). \end{cases}$

We repeat the two "firstly" equations of §34 by saying that

(8) $\begin{cases} \text{if } (B,\tilde{B},\hat{B},\hat{C},b,c) \in \text{supt}_6(\ell(e)) \text{ is such that } b < B \\ \text{then } e^*(B,\tilde{B},\hat{B},\hat{C},b,c) = 0 = e^*((B,\tilde{B},\hat{B},\hat{C},b,c)) \end{cases}$

and we repeat the "decreasing induction" equation of §34 by saying that

$$(9)\quad\begin{cases}\text{if } (B,\tilde{B},\hat{B},\hat{C},b,c) \in \mathrm{supt}_6(\ell(e)) \text{ is such that} \\ B \le \tilde{B} \le \hat{B} \quad\text{and}\quad B \le b < \tilde{B} \\ \text{then}\quad e^*(B,\tilde{B},\hat{B},\hat{C},b,c) \\ \qquad = \sum_{\substack{b+1 \le b' \le o(\ell(e)) \\ 0 \le c' \le b'(\ell(e))}} e^*(B,\tilde{B},\hat{B},\hat{C},b',c')e(b+1,b',c')\end{cases}$$

and we repeat the "finally" equation of §34 by saying that

$$(10)\quad\begin{cases}\text{if } (B,\tilde{B},\hat{B},\hat{C},b,c) \in \mathrm{supt}_6(\ell(e)) \text{ is such that } B \le b \\ \text{then}\quad e^*((B,\tilde{B},\hat{B},\hat{C},b,c)) \\ \qquad = \sum_{\substack{b \le b' \le o(\ell(e)) \\ 0 \le c' \le b'(\ell(e))}} e^*(B,\tilde{B},\hat{B},\hat{C},b',c')e(b,b',c').\end{cases}$$

By (8) and (10) we see that

$$(11)\quad \mathrm{supt}((e^*)) \subset \mathrm{supt}(e^*).$$

By the first and the second and the fifth and the sixth of the "begin with" equations of §34 we see that

$$(12.1)\begin{cases}\text{if } (B,\tilde{B},\hat{B},\hat{C},b,c) \in \mathrm{supt}_6(\ell(e)) \text{ is such that} \\ \hat{C} = 0 \quad\text{and}\quad \tilde{B} \le b \quad\text{and} \\ \text{either } B \le \tilde{B} < \hat{B} \quad\text{or}\quad B \le \tilde{B} = \hat{B} \le o(\ell(e))-2 \\ \text{then}\quad e^*(B,\tilde{B},\hat{B},\hat{C},b,c) = 0.\end{cases}$$

In view of (9) and (12.1), by decreasing induction on b we see that

$$(12.2)\begin{cases} \text{if } (B,\tilde{B},\hat{B},\hat{C},b,c) \in \mathrm{supt}_6(\ell(e)) \text{ is such that} \\ \hat{C} = 0 \text{ and } B \le b < \tilde{B} \text{ and} \\ \text{either } B \le \tilde{B} < \hat{B} \text{ or } B \le \tilde{B} = \hat{B} \le o(\ell(e))-2 \\ \text{then } e^*(B,\tilde{B},\hat{B},\hat{C},b,c) = 0 . \end{cases}$$

By (8), (12.1) and (12.2) we see that

$$(12.3)\begin{cases} \text{if } (B,\tilde{B},\hat{B},\hat{C},b,c) \in \mathrm{supt}_6(\ell(e)) \text{ is such that} \\ \hat{C} = 0 \text{ and either } B \le \tilde{B} < \hat{B} \text{ or } B \le \tilde{B} = \hat{B} \le o(\ell(e))-2 \\ \text{then } e^*(B,\tilde{B},\hat{B},\hat{C},b,c) = 0 . \end{cases}$$

In view of (11) and (12.3), by the two "secondly" equations of §34 we see that

$$(12)\quad\begin{cases} \mathrm{supt}(e^*) \cup \mathrm{supt}((e^{**})) \\ \subset \{(B,\tilde{B},\hat{B},\hat{C}) \in \mathrm{supt}_4(\ell(e)) : \text{ either } \hat{C} \neq 0 < B \le \tilde{B} \le \hat{B} \\ \qquad\qquad\qquad\qquad\qquad\qquad \text{ or } \hat{C} = 0 < B \le \tilde{B} = B \ge o(\ell(e))-1\}. \end{cases}$$

In view of (10) and (12), by the first of the "begin with" equations of §34 we see that

$$(13)\quad\begin{cases} \text{if } (B,\tilde{B},\hat{B},\hat{C},b,c) \in \mathrm{supt}_6(\ell(e)) \text{ is such that } b > \hat{B} \\ \text{then } e^*(B,\tilde{B},\hat{B},\hat{C},b,c) = 0 = e^*((B,\tilde{B},\hat{B},\hat{C},b,c)). \end{cases}$$

By the first four of the "begin with" equations of §34 we see that

$$(14)\quad \begin{cases} \text{for any } (B,\widetilde{B},\hat{B},\hat{C},b,c) \in \text{supt}_6(\ell(e)) \quad \text{with} \\ \hat{C} \neq 0 < B \leq \widetilde{B} \leq \hat{B} \quad \text{and} \quad \widetilde{B} \leq b \quad \text{we have} \\ e^*(B,\widetilde{B},\hat{B},\hat{C},b,c) = \begin{cases} 0 & \text{if } (b,c) \neq (\hat{B},\hat{C}) \\ 1 & \text{if } (b,c) = (\hat{B},\hat{C}) \end{cases} \end{cases}$$

and by the first and the seventh of the "begin with" equations of §34 we see that

$$(15)\quad \begin{cases} \text{for any } (B,\widetilde{B},\hat{B},\hat{C},b,c) \in \text{supt}_6(\ell(e)) \quad \text{with} \\ \hat{C} = 0 < B \leq \widetilde{B} = \hat{B} \geq o(\ell(e))-1 \quad \text{and} \quad \widetilde{B} \leq b \quad \text{we have} \\ e^*(B,\widetilde{B},\hat{B},\hat{C},b,c) = \begin{cases} 0 & \text{if } b \neq \hat{B} \\ 1 & \text{if } b = \hat{B} \end{cases} . \end{cases}$$

By (8), (9), (12), (14) and (15) we see that

$$(16)\quad \begin{cases} \text{for any } (B,\widetilde{B},\hat{B},\hat{C},b,c) \in \text{supt}_6(\ell(e)) \\ \text{we have:} \quad e^*(B,\widetilde{B},\hat{B},\hat{C},b,c) \neq e^*(B,\widetilde{B},\hat{B},\hat{C},b,0) \\ \qquad \Leftrightarrow e^*(B,\widetilde{B},\hat{B},\hat{C},b,c) \neq 0 = e^*(B,\widetilde{B},\hat{B},\hat{C},b,0) \\ \qquad \Leftrightarrow B \leq \widetilde{B} \leq \hat{B} = b \quad \text{and} \quad c = \hat{C} \neq 0. \end{cases}$$

By (8) and (10) we see that

$$(17)\quad \begin{cases} \text{for any } (B,\widetilde{B},\hat{B},\hat{C},b,c) \in \text{supt}_6(\ell(e)) \\ \text{we have: } e^*((B,\widetilde{B},\hat{B},\hat{C},b,c)) = e^*((B,\widetilde{B},\hat{B},\hat{C},b,0)). \end{cases}$$

By (12), (14) and (15) we see that

$$\text{(18.1)}\quad \begin{cases} \text{if } (B,\tilde{B},\hat{B},\hat{C},b,c) \in \operatorname{supt}_6(\ell(e)) \text{ is such that} \\ B+1 \le \tilde{B} \le \hat{B} \text{ and } \tilde{B} \le b \\ \text{then } e^*(B,\tilde{B},\hat{B},\hat{C},b,c) = e^*(B+1,\tilde{B},\hat{B},\hat{C},b,c). \end{cases}$$

In view of (9) and (18.1), by decreasing induction on b we see that

$$\text{(18.2)}\quad \begin{cases} \text{if } (B,\tilde{B},\hat{B},\hat{C},b,c) \in \operatorname{supt}_6(\ell(e)) \text{ is such that} \\ B+1 \le \tilde{B} \le \hat{B} \text{ and } B+1 \le b \\ \text{then } e^*(B,\tilde{B},\hat{B},\hat{C},b,c) = e^*(B+1,\tilde{B},\hat{B},\hat{C},b,c). \end{cases}$$

By (8) and (18.2) we see that

$$\text{(18)}\quad \begin{cases} \text{if } (B,\tilde{B},\hat{B},\hat{C},b,c) \in \operatorname{supt}_6(\ell(e)) \text{ is such that} \\ B+1 \le \tilde{B} \le \hat{B} \text{ and } b \ne B \\ \text{then } e^*(B,\tilde{B},\hat{B},\hat{C},b,c) = e^*(B+1,\tilde{B},\hat{B},\hat{C},b,c). \end{cases}$$

By (14) we see that

$$\text{(19.1)}\quad \begin{cases} \text{if } (B,\tilde{B},\hat{B},\hat{C}) \in \operatorname{supt}_4(\ell(e)) \text{ is such that} \\ \hat{C} \ne 0 < B = \tilde{B} \le \hat{B} \\ \text{then } e^*(B,\tilde{B},\hat{B},\hat{C},B,0) = 0 . \end{cases}$$

Obviously

$$\text{(19.2)}\quad \begin{cases} \text{if } (B,\tilde{B},\hat{B},\hat{C}) \in \operatorname{supt}_4(\ell(e)) \text{ is such that } B \le o(\ell(e))-2 \\ \text{and either } \hat{C} \ne 0 < B = \tilde{B} \le \hat{B} \text{ or } \hat{C} = 0 < B = \tilde{B} = \hat{B} \ge o(\ell(e))-1 \\ \text{then } \hat{C} \ne 0 < B = \tilde{B} \le \hat{B}. \end{cases}$$

By (19.1) and (19.2) we see that

$$(19.3)\quad \begin{cases} \text{if } (B,\tilde{B},\hat{B},\hat{C}) \in \text{supt}_4(\ell(e)) \text{ is such that } B \le o(\ell(e))-2 \\ \text{and either } \hat{C} \ne 0 < B = \tilde{B} \le \hat{B} \quad \text{or} \quad \hat{C} = 0 \le B = \tilde{B} = \hat{B} \ge o(\ell(e))-1 \\ \text{then } e^*(B,\tilde{B},\hat{B},\hat{C},B,0) = 0 \,. \end{cases}$$

By (12) we know that

$$(19.4)\quad \begin{cases} \text{if } (B,\tilde{B},\hat{B},\hat{C}) \in \text{supt}_4(\ell(e)) \text{ is such that} \\ e^*(B,\tilde{B},\hat{B},\hat{C},B,0) \ne 0 \\ \text{then either } \hat{C} \ne 0 < B \le \tilde{B} \le \hat{B} \quad \text{or} \quad \hat{C} = 0 < B \le \tilde{B} = \hat{B} \ge o(\ell(e))-1. \end{cases}$$

By (19.3) and (19.4) it follows that

$$(19.5)\quad \begin{cases} \text{if } (B,\tilde{B},\hat{B},\hat{C}) \in \text{supt}_4(\ell(e)) \text{ is such that } B \le o(\ell(e))-2 \\ \text{and } e^*(B,\tilde{B},\hat{B},\hat{C},B,0) \ne 0 \\ \text{then } B+1 \le \tilde{B} \le \hat{B}. \end{cases}$$

By (11) and (12) we also see that

$$(19.6)\quad \begin{cases} \text{if } (B,\tilde{B},\hat{B},\hat{C}) \in \text{supt}_4(\ell(e)) \text{ is such that } B \le o(\ell(e))-1 \\ \text{and } e^*((B+1,\tilde{B},\hat{B},\hat{C},B+1,0)) \ne 0 \\ \text{then } B+1 \le \tilde{B} \le \hat{B} \,. \end{cases}$$

By (19.5) and (19.6) it follows that

$$(19)\quad \begin{cases} \text{if } (B,\tilde{B},\hat{B},\hat{C}) \in \text{supt}_4(\ell(e)) \text{ is such that } B \le o(\ell(e))-2 \\ \text{and either } e^*(B,\tilde{B},\hat{B},\hat{C},B,0) \ne 0 \text{ or } e^*((B+1,\tilde{B},\hat{B},\hat{C},B+1,0)) \ne 0 \\ \text{then } B+1 \le \tilde{B} \le \hat{B} \,. \end{cases}$$

By (10) we see that

$$(20.1)\quad \begin{cases} \text{if } (B,\tilde{B},\hat{B},\hat{C}) \in \text{supt}_4(\ell(e)) \text{ is such that } B+1 \le \tilde{B} \le \hat{B} \\ \text{then } e^*((B+1,\tilde{B},\hat{B},\hat{C},B+1,0)) \\ \qquad = \sum_{\substack{B+1\le b'\le o(\ell(e)) \\ 0\le c'\le b'(\ell(e))}} e^*(B+1,\tilde{B},\hat{B},\hat{C},b',c')e(B+1,b',c'). \end{cases}$$

By (18) and (20.1) we see that

$$(20.2)\quad \begin{cases} \text{if } (B,\tilde{B},\hat{B},\hat{C}) \in \text{supt}_4(\ell(e)) \text{ is such that } B+1 \le \tilde{B} \le \hat{B} \\ \text{then } e^*((B+1,\tilde{B},\hat{B},\hat{C},B+1,0)) \\ \qquad = \sum_{\substack{B+1\le b'\le o(\ell(e)) \\ 0\le c'\le b'(\ell(e))}} e^*(B,\tilde{B},\hat{B},\hat{C},b',c')e(B+1,b',c') . \end{cases}$$

By (9) and (20.2) we see that

$$(20.3)\quad \begin{cases} \text{if } (B,\tilde{B},\hat{B},\hat{C}) \in \text{supt}_4(\ell(e)) \text{ is such that } B+1 \le \tilde{B} \le \hat{B} \\ \text{then } e^*((B+1,\tilde{B},\hat{B},\hat{C},B+1,0)) = e^*(B,\tilde{B},\hat{B},\hat{C},B,0). \end{cases}$$

By (19) and (20.3) it follows that

$$(20)\quad \begin{cases} \text{if } (B,\tilde{B},\hat{B},\hat{C}) \in \text{supt}_4(\ell(e)) \text{ is such that } B \le o(\ell(e))-2 \\ \text{then } e^*(B,\tilde{B},\hat{B},\hat{C},B,0) = e^*((B+1,\tilde{B},\hat{B},\hat{C},B+1,0)). \end{cases}$$

By (3), (4), (5), (7), (9) and (12) we see that

$$(21)\quad \begin{cases} \text{if } (B,\widetilde{B},\hat{B},\hat{C},b,c) \in \mathrm{supt}_6(\ell(e)) \text{ is such that } B \le b < \widetilde{B} \\ \text{then } e^*(B,\widetilde{B},\hat{B},\hat{C},b,c) = e[b+1]e^*(B,\widetilde{B},\hat{B},\hat{C},b+1,0) \\ \qquad\qquad +\mathrm{inpo}(e(b+1),e^*(B,\widetilde{B},\hat{B},\hat{C})). \end{cases}$$

By (3), (4), (5), (7) and (10) we see that

$$(22)\quad \begin{cases} \text{if } (B,\widetilde{B},\hat{B},\hat{C},b,c) \in \mathrm{supt}_6(\ell(e)) \text{ is such that } B \le b \\ \text{then } e^*((B,\widetilde{B},\hat{B},\hat{C},b,c)) = e[b]e^*(B,\widetilde{B},\hat{B},\hat{C},b,0) \\ \qquad\qquad + \mathrm{inpo}(e(b),e^*(B,\widetilde{B},\hat{B},\hat{C})). \end{cases}$$

By (3), (9), (10) and (14) we see that

$$(23)\quad \begin{cases} \text{for any } (B,\widetilde{B},\hat{B},\hat{C},b,c) \in \mathrm{supt}_6(\ell(e)) \text{ with} \\ \hat{C} \ne 0 < B \le \widetilde{B} \le \hat{B} \text{ we have} \\ e^*(B,\widetilde{B},\hat{B},\hat{C},b,c) = \begin{cases} 0 & \text{if } b \ge \widetilde{B} \text{ and } (b,c) \ne (\hat{B},\hat{C}) \\ 1 & \text{if } (b,c) = (\hat{B},\hat{C}) \\ e(\widetilde{B},\hat{B},\hat{C}) & \text{if } B \le b = \widetilde{B}-1 \end{cases} \\ \text{and } e^*((B,\widetilde{B},\hat{B},\hat{C},b,c)) = e(b,\hat{B},\hat{C}) \quad \text{if } b \ge \widetilde{B}. \end{cases}$$

By (9), (10) and (15) we see that

$$(24)\quad \begin{cases} \text{for any } (B,\widetilde{B},\hat{B},\hat{C},b,c) \in \mathrm{supt}_6(\ell(e)) \text{ with} \\ \hat{C} = 0 < B \le \widetilde{B} = \hat{B} = o(\ell(e))-1 \text{ we have} \\ e^*(B,\widetilde{B},\hat{B},\hat{C},b,c) = \begin{cases} 0 & \text{if } b > \widetilde{B} \\ 1 & \text{if } b = \widetilde{B} \\ \sum_{0 \le \widetilde{C} \le \widetilde{B}(\ell(e))} e(\widetilde{B},\widetilde{B},\widetilde{C}) & \text{if } B \le b = \widetilde{B}-1 \end{cases} \\ \text{and } e^*((B,\widetilde{B},\hat{B},\hat{C},b,c)) = \begin{cases} 0 & \text{if } b > \widetilde{B} \\ \sum_{0 \le \widetilde{C} \le \widetilde{B}(\ell(e))} e(\widetilde{B},\widetilde{B},\widetilde{C}) & \text{if } b = \widetilde{B} \end{cases} \end{cases}$$

By (3), (6), (9), (10) and (15) we see that

$$(25)\quad \begin{cases} \text{for any } (B,\tilde{B},\hat{B},\hat{C},b,c) \in \operatorname{supt}_6(\ell(e)) \text{ with} \\ \hat{C} = 0 < B \le \tilde{B} = \hat{B} = o(\ell(e)) \quad \text{we have} \\ e^*(B,\tilde{B},\hat{B},\hat{C},b,c) = \begin{cases} 1 & \text{if } \ b = \tilde{B} \\ e[[\tilde{B}]] \neq 0 & \text{if } \ B \le b = \tilde{B} - 1 \end{cases} \\ \text{and } \ e^*((B,\tilde{B},\hat{B},\hat{C},b,c)) = e[[\tilde{B}]] \neq 0 \ \text{ if } \ b = \tilde{B}. \end{cases}$$

In view of (6), (9), (10) and (25), by decreasing induction on b we see that

$$(26)\quad \begin{cases} \text{for any } (B,\tilde{B},\hat{B},\hat{C},b,c) \in \operatorname{supt}_6(\ell(e)) \text{ with} \\ \hat{C} = 0 < B \le \tilde{B} = \hat{B} = o(\ell(e)) \ \text{ and } \ B \le b \ \text{ we have} \\ e^*(B,\tilde{B},\hat{B},\hat{C},b,c) \neq 0 \neq e^*((B,\tilde{B},\hat{B},\hat{C},b,c)) . \end{cases}$$

Recall that

$$(31)\quad \begin{cases} \text{for all } (B,\tilde{B},\hat{B},\hat{C},b,c) \in \operatorname{supt}_6(\ell(e)) \text{ we have} \\ e^{**}(B,\tilde{B},\hat{B},\hat{C},b,c) = \begin{cases} \dfrac{e^*(B,\tilde{B},\hat{B},\hat{C},b,c)}{e^*(B,\tilde{B},\hat{B},\hat{C},B,0)} & \text{if } e^*(B,\tilde{B},\hat{B},\hat{C},B,0) \neq 0 \\ 0 & \text{if } \ e^*(B,\tilde{B},\hat{B},\hat{C},B,0) = 0 \end{cases} \end{cases}$$

and

$$(32)\quad \begin{cases} \text{for all } (B,\widetilde{B},\hat{B},\hat{C},b,c) \in \mathrm{supt}_6(\ell(e)) \text{ we have} \\ e^{**}((B,\widetilde{B},\hat{B},\hat{C},b,c)) = \begin{cases} \dfrac{e^*(B,\widetilde{B},\hat{B},\hat{C},b,c)}{e^*((B,\widetilde{B},\hat{B},\hat{C},B,0))} & \text{if } e^*((B,\widetilde{B},\hat{B},\hat{C},B,0)) \neq 0 \\ 0 \quad \text{if } e^*((B,\widetilde{B},\hat{B},\hat{C},B,0)) = 0 . \end{cases} \end{cases}$$

By (8), (31) and (32) we see that

$$(33)\quad \begin{cases} \text{if } (B,\widetilde{B},\hat{B},\hat{C},b,c) \in \mathrm{supt}_6(\ell(e)) \text{ is such that } b < B \\ \text{then } e^{**}(B,\widetilde{B},\hat{B},\hat{C},b,c) = 0 = e^{**}((B,\widetilde{B},\hat{B},\hat{C},b,c)). \end{cases}$$

By (26), (31), and (32) we see that

$$(34)\quad \begin{cases} \text{for all } (B,b,c) \in \mathrm{supt}_3(\ell(e)) \text{ with } b \geq B \text{ we have} \\ e^{**}(B,o(\ell(e)),o(\ell(e)),0,b,c) \neq 0 \neq e^{**}((B,o(\ell(e),o(\ell(e)),0,b,c)) \\ \text{and so in particular for every } B \in [1,o(\ell(e))] \text{ we have} \\ (o(\ell(e)),o(\ell(e)),0) \in \mathrm{supt}(e^{**},B) \cap \mathrm{supt}((e^{**},B)) . \end{cases}$$

By (33) and (34) it follows that

(35) e^{**} is a scale .

By (13), (31) and (32) we see that

$$(36)\quad \begin{cases} \text{if } (B,\widetilde{B},\hat{B},\hat{C},b,c) \in \mathrm{supt}_6(\ell(e)) \text{ is such that } b > \hat{B} \\ \text{then } e^{**}(B,\widetilde{B},\hat{B},\hat{C},b,c) = 0 = e^{**}((B,\widetilde{B},\hat{B},\hat{C},b,c)). \end{cases}$$

By (16) and (31) we see that

$$(37)\quad \begin{cases} \text{given any } (B,\widetilde{B},\hat{B},\hat{C},b,c) \in \mathrm{supt}_6(\ell(e)) \\ \text{we have: } e^{**}(B,\widetilde{B},\hat{B},\hat{C},b,c) \neq e^{**}\ (B,\widetilde{B},\hat{B},\hat{C},b,0) \\ \qquad \Leftrightarrow e^{**}\ (B,\widetilde{B},\hat{B},\hat{C},b,c) \neq 0 = e^{**}\ (B,\widetilde{B},\hat{B},\hat{C},b,0)\ . \end{cases}$$

By (16) and (32) we see that

$$(38)\quad \begin{cases} \text{given any } (B,\widetilde{B},\hat{B},\hat{C},b,c) \in \mathrm{supt}_6(\ell(e)) \\ \text{we have: } e^{**}((B,\widetilde{B},\hat{B},\hat{C},b,c)) \neq e^{**}((B,\widetilde{B},\hat{B},\hat{C},b,0)) \\ \qquad \Leftrightarrow e^{**}((B,\widetilde{B},\hat{B},\hat{C},b,c)) \neq 0 = e^{**}((B,\widetilde{B},\hat{B},\hat{C},b,0)). \end{cases}$$

By (31) and (32) we see that

$$(39)\quad \begin{cases} \text{for every } B \in [1,o(\ell(e))] \text{ we have} \\ \mathrm{supt}(e^{**},B) \cup \mathrm{supt}((e^{**},B)) \\ \subset \mathrm{supt}(e^{*},B) \cup \mathrm{supt}((e^{*},B)). \end{cases}$$

By (11) and (39) we see that

$$(40)\quad \begin{cases} \text{for every } B \in [1,o(\ell(e))] \text{ we have} \\ \mathrm{supt}(e^{**},B) \cup \mathrm{supt}((e^{**},B)) \\ \subset \mathrm{supt}(e^{*},B). \end{cases}$$

By (12) and (39) we see that

$$(41)\quad \begin{cases} \text{for every } B \in [1,o(\ell(e))] \text{ we have} \\ \mathrm{supt}(e^{**},B) \cup \mathrm{supt}((e^{**},B)) \\ \subset \{(\tilde{B},\hat{B},\hat{C}) \in \mathrm{supt}_3(\ell(e)) : \text{either } \hat{C} \neq 0 < B \leq \tilde{B} \leq \hat{B} \\ \qquad\qquad \text{or } \hat{C} = 0 < B \leq \tilde{B} = \hat{B} \geq o(\ell(e))-1\}. \end{cases}$$

By (31) we see that

$$(42)\quad \begin{cases} \text{for every } B \in [1,o(\ell(e))] \text{ we have} \\ \mathrm{supt}(e^{**},B) \\ = \{(\tilde{B},\hat{B},\hat{C}) \in \mathrm{supt}_3(\ell(e)) : e^{**}(B,\tilde{B},\hat{B},\hat{C},B,0) \neq 0\} \\ = \{(\tilde{B},\hat{B},\hat{C}) \in \mathrm{supt}_3(\ell(e)) : e^{**}(B,\tilde{B},\hat{B},\hat{C},B,0) = 1\}\ . \end{cases}$$

By (31) we see that

$$(43)\quad \begin{cases} \text{for every } B \in [1,o(\ell(e))] \text{ we have} \\ \mathrm{supt}(e^{**},B) \\ = \{(\tilde{B},\hat{B},\hat{C}) \in \mathrm{supt}_3(\ell(e)) : e^{*}(B,\tilde{B},\hat{B},\hat{C},B,0) \neq 0\}\ . \end{cases}$$

By (32) we see that

$$(44)\quad \begin{cases} \text{for every } B \in [1,o(\ell(E))] \text{ we have} \\ \mathrm{supt}((e^{**},B)) \\ = \{(\tilde{B},\hat{B},\hat{C}) \in \mathrm{supt}(e^{*},B) : e^{**}((B,\tilde{B},\hat{B},\hat{C},B,0)) \neq 0\}\ . \end{cases}$$

By (11) and (44) we see that

$$(45)\quad \begin{cases} \text{for every } B \in [1,o(\ell(e))] \text{ we have} \\ \text{supt}((e^{**},B)) \\ = \{(\tilde{B},\hat{B},\hat{C}) \in \text{supt}_3(\ell(e)) : e^*((B,\tilde{B},\hat{B},\hat{C},B,0)) \neq 0\} . \end{cases}$$

By (32) we see that

$$(46)\quad \begin{cases} \text{for any } (B,\tilde{B},\hat{B},\hat{C}) \in \text{supt}_4(\ell(e)) \\ \text{we have: } e^{**}((B,\tilde{B},\hat{B},\hat{C},B,0)) \neq 0 \\ \qquad \Leftrightarrow e^*(B,\tilde{B},\hat{B},\hat{C},B,0) \neq 0 \neq e^*((B,\tilde{B},\hat{B},\hat{C},B,0)). \end{cases}$$

By (43) and (45) we see that

$$(47)\quad \begin{cases} \text{for every } B \in [1,o(\ell(e))] \text{ we have} \\ \text{supt}(e^{**},B) \cap \text{supt}((e^{**},B)) \\ = \{(\tilde{B},\hat{B},\hat{C}) \in \text{supt}_3(\ell(e)) : e^*(B,\tilde{B},\hat{B},\hat{C},B,0) \neq 0 \text{ and} \\ \qquad e^*((B,\tilde{B},\hat{B},\hat{C},B,0)) \neq 0\} . \end{cases}$$

By (46) and (47) we see that

$$(48)\quad \begin{cases} \text{for every } B \in [1,o(\ell(e))] \text{ we have} \\ \text{supt}(e^{**},B) \cap \text{supt}((e^{**},B)) \\ = \{(\tilde{B},\hat{B},\hat{C}) \in \text{supt}_3(\ell(e)) : e^{**}((B,\tilde{B},\hat{B},\hat{C},B,0)) \neq 0\} . \end{cases}$$

By (2), (23), (24), (26), (31), (33), (41) and (43) we see that

$$(49)\quad \begin{cases} \text{if } 1 \le B = o(\ell(e))-1 \\ \text{then } \mathrm{supt}(e^{**},B) = \{(\tilde{B},\tilde{B},0): \tilde{B} \in [B,B+1]\} \\ \text{and for any } (b,c) \in \mathrm{supt}_2(\ell(e)) \text{ we have} \\ e^{**}(B,\tilde{B},\tilde{B},0,b,c) \begin{cases} = 0 & \text{if } B = \tilde{B} \neq b \\ = 1 & \text{if } B = \tilde{B} = b \\ = 0 & \text{if } B+1 = \tilde{B} \neq b \neq B \\ = 1 & \text{if } B+1 = \tilde{B} \text{ and } b = B \\ \neq 0 & \text{if } B+1 = \tilde{B} = b. \end{cases} \end{cases}$$

By (49) we see that

$$(50)\quad \begin{cases} \text{if } 1 \le B = o(\ell(e))-1 \quad \text{and} \quad B' \in [1,B] \quad \text{and } u \in Q \\ \text{then} \\ Q(e^{**}(B) \ge u) \cap Q(\ell(e),B') = \{j \in Q(\ell(e),B'): \mathrm{abs}(j(B)) \ge u\} \\ \text{and} \\ Q(e^{**}(B) = u) \cap Q(\ell(e),B') = \{j \in Q(\ell(e),[B',B]): \mathrm{abs}(j(B)) = u\}. \end{cases}$$

By (18), (19), (20), (31), (32), (37), (38), (43) and (45) we see that

$$(51)\quad \text{if } B \in [1,o(\ell(e))-2] \text{ then } \mathrm{supt}(e^{**},B) = \mathrm{supt}((e^{**},B+1))$$

and that

$$(52)\quad \begin{cases} \text{if } B \in [1,o(\ell(e))-2] \text{ and } (\tilde{B},\hat{B},\hat{C}) \in \mathrm{supt}(e^{**},B) \\ \text{then for any } (b,c) \in \mathrm{supt}_2(\ell(e)) \text{ we have} \\ e^{**}(B,\tilde{B},\hat{B},\hat{C},b,c) = \begin{cases} e^{**}((B+1,\tilde{B},\hat{B},\hat{C},b,c)) & \text{if } b \neq B \\ 1 \quad \text{if} \quad b = B\,. \end{cases} \end{cases}$$

By (52) we see that

$$(53)\quad \begin{cases} \text{if } B \in [1,o(\ell(e))-2] \text{ and } (\tilde{B},\hat{B},\hat{C}) \in \text{supt}(e^{**},B) \text{ then} \\ \text{inpo}(i,e^{**}(B,\tilde{B},\hat{B},\hat{C})) \\ = \begin{cases} \text{inpo}(i,e^{**}((B+1,\tilde{B},\hat{B},\hat{C}))) \text{ for all } i \in Q(\ell(e),B+1) \\ \text{abs}(i) \text{ for all } i \in Q(\ell(e),1,B) \ . \end{cases} \end{cases}$$

Obviously

$$(54)\quad \begin{cases} \text{for any } B \in [1,o(\ell(e))] \text{ we have} \\ Q(\ell(e),B) = \{i+j\colon i \in Q(\ell(e),1,B) \text{ and } j \in Q(\ell(e),B+1)\}. \end{cases}$$

By (34), (51), (53) and (54) we see that

$$(55)\quad \begin{cases} \text{if } B \in [1,o(\ell(e))-2] \text{ and } u \in Q \text{ and } P \in \{=,\geq\} \text{ then} \\ Q(e^{**}(B)Pu) \cap Q(\ell(e),B) \\ = \bigcup\limits_{\substack{v\in Q \text{ and } w\in Q \\ \text{with } v+w=u}} \left\{ i+j\colon \begin{array}{l} j \in Q(e^{**}((B+1))Pw) \cap Q(\ell(e),B+1) \\ \text{and } i \in Q(\ell(e)Pv,1,B) \end{array} \right\}. \end{cases}$$

By (22) we see that

$$(61)\quad \begin{cases} \text{for every } (B,\tilde{B},\hat{B},\hat{C}) \in \text{supt}_4(\ell(e)) \text{ we have} \\ e^*((B,\tilde{B},\hat{B},\hat{C},B,0)) = e[B]e^*(B,\tilde{B},\hat{B},\hat{C},B,0) \\ \qquad\qquad + \text{inpo}(e(B),e^*(B,\tilde{B},\hat{B},\hat{C})) \ . \end{cases}$$

By (43), (45) and (61) we see that

$$(62)\quad \begin{cases} \text{if } (B,\tilde{B},\hat{B},\hat{C}) \in \text{supt}_4(\ell(e)) \text{ is such that} \\ (\tilde{B},\hat{B},\hat{C}) \in \text{supt}(e^*,B)\setminus\text{supt}((e^{**},B)) \\ \text{then } e[B] = 0 \ . \end{cases}$$

By (32), (43), (45) and (61) we see that

$$
(63)\quad \begin{cases} \text{if } (\widetilde{B},\hat{B},\hat{B},C) \in \mathrm{supt}_4(\ell(e)) \text{ is such} \\ (\widetilde{B},\hat{B},\hat{C}) \in \mathrm{supt}((e^{**},B))\setminus \mathrm{supt}(e^{*},B) \\ \text{then } \mathrm{inpo}(e(B),e^{**}((B,\widetilde{B},\hat{B},\hat{C}))) = 1 . \end{cases}
$$

By (32), (46) and (61) we see that

$$
(64)\quad \begin{cases} \text{if } (B,\widetilde{B},\hat{B},\hat{C}) \in \mathrm{supt}_4(\ell(e)) \text{ is such that} \\ e^{**}((B,\widetilde{B},\hat{B},\hat{C},B,0)) \neq 0 \\ \text{then } e[B]e^{**}((B,\widetilde{B},\hat{B},\hat{C},B,0)) + \mathrm{inpo}(e(B),e^{**}((B,\widetilde{B},\hat{B},\hat{C}))) = 1. \end{cases}
$$

By (31) and (32) we see that

$$
(65)\quad \begin{cases} \text{if } (B,\widetilde{B},\hat{B},\hat{C}) \in \mathrm{supt}_4(\ell(e)) \text{ is such that} \\ e^{**}((B,\widetilde{B},\hat{B},\hat{C},B,0)) \neq 0 \\ \text{then for all } (b,c) \in \mathrm{supt}_2(\ell(e)) \text{ we have} \\ e^{**}((B,\widetilde{B},\hat{B},\hat{C},b,c)) \\ = e^{**}((B,\widetilde{B},\hat{B},\hat{C},B,0))e^{**}(B,\widetilde{B},\hat{B},\hat{C},b,c). \end{cases}
$$

Upon taking $B = \widetilde{B}$ in (23), in view of (3), (32), (33) and (45) we see that

$$
(66)\quad \begin{cases} \text{if } B \in [1,o(\ell(e))] \text{ and } (\hat{B},\hat{C}) \in \mathrm{supt}_2(\ell(e)) \\ \text{are such that } \hat{C} \neq 0 \neq e(B,\hat{B},\hat{C}) \\ \text{then } B \le \hat{B} \text{ and } (B,\hat{B},\hat{C}) \in \mathrm{supt}((e^{**},B)) \\ \text{and for all } (b,c) \in \mathrm{supt}_2(\ell(e)) \text{ we have} \\ e^{**}((B,B,\hat{B},\hat{C},b,c)) = \begin{cases} 0 \text{ if } (b,c) \neq (\hat{B},\hat{C}) \\ e(B,\hat{B},\hat{C})^{-1} \text{ if } (b,c) = (\hat{B},\hat{C}). \end{cases} \end{cases}
$$

By (66) we see that

$$
(67)\quad \begin{cases} \text{if } B \in [1, o(\ell(e))] \text{ and } (\hat{B},\hat{C}) \in \mathrm{supt}(\ell(e)) \\ \text{are such that } e(B,\hat{B},\hat{C}) \neq 0 \\ \text{then } (B,\hat{B},\hat{C}) \in \mathrm{supt}((e^{**},B)) \\ \text{and for all } j \in Q(\ell(e)) \text{ we have} \\ \mathrm{inpo}(j, e^{**}((B,B,\hat{B},\hat{C}))) = j(\hat{B},\hat{C})e(B,\hat{B},\hat{C})^{-1} . \end{cases}
$$

By (67) it follows that

$$
(68)\quad \begin{cases} \text{if } B \in [1, o(\ell(e))] \text{ and } u \in Q \text{ and } j \in Q(\ell(e)) \\ \text{are such that } j \in Q(e^{**}((B)) \geq u) \\ \text{then for all } (\hat{B},\hat{C}) \in \mathrm{supt}(\ell(e)) \text{ with } e(B,\hat{B},\hat{C}) \neq 0 \\ \text{we have } j(\hat{B},\hat{C})e(B,\hat{B},\hat{C})^{-1} \geq u \\ \text{and hence we have } j(\hat{B},\hat{C}) \geq ue(B,\hat{B},\hat{C}). \end{cases}
$$

By (68) we see that

$$
(69)\quad \begin{cases} \text{given any } B \in [1, o(\ell(e))] \text{ and} \\ \text{given any } u \in Q \text{ and } j \in Q(e^{**}((B)) \geq u) \\ \text{there exists } i \in Q(\ell(e)) \text{ such that } j = i + ue(B). \end{cases}
$$

By (65) we see that

$$
(70)\quad \begin{cases} \text{if } (B,\tilde{B},\hat{B},\hat{C}) \in \mathrm{supt}_4(\ell(e)) \text{ is such that} \\ e^{**}((B,\tilde{B},\hat{B},\hat{C},B,0)) \neq 0 \\ \text{then for all } i \in Q(\ell(e)) \text{ we have:} \\ e^{**}((B,\tilde{B},\hat{B},\hat{C},B,0))\mathrm{inpo}(i,e^{**}(B,\tilde{B},\hat{B},\hat{C})) \\ = \mathrm{inpo}(i,e^{**}((B,\tilde{B},\hat{B},\hat{C}))) . \end{cases}
$$

For any $P \in \{=,\geq\}$ and $A_1 \in Q$ and $A_2 \in Q$ and $0 \neq A \in Q$ we clearly have $[A_1PA_2 \Leftrightarrow AA_1PAA_2]$, and hence by (70) we see that

$$
(71)\quad \begin{cases}
\text{if } (B,\tilde{B},\hat{B},\hat{C}) \in \mathrm{supt}_4(\ell(e)) \text{ is such that} \\
e^{**}((B,\tilde{B},\hat{B},\hat{C},B,0)) \neq 0 \\
\text{then for any } i \in Q(\ell(e)) \text{ and } u \in Q \text{ and } P \in \{=,\geq\} \\
\text{we have:} \\
\quad \begin{cases} \mathrm{inpo}(i,e^{**}(B,\tilde{B},\hat{B},\hat{C})) \\ \mathrm{Pue}[B] \end{cases} \\
\quad \Leftrightarrow \begin{cases} \mathrm{inpo}(i,e^{**}((B,\tilde{B},\hat{B},\hat{C}))) \\ \mathrm{Pue}[B]e^{**}((B,\tilde{B},\hat{B},\hat{C},B,0)). \end{cases}
\end{cases}
$$

For any $P \in \{=,\geq\}$ and $A_1 \in Q$ and $A_2 \in Q$ and $A \in Q$ we clearly have $[A_1PA_2 \Leftrightarrow (A_1+A)P(A_2+A)]$, and hence we see that

$$
(72)\quad \begin{cases}
\text{for any } (B,\tilde{B},\hat{B},\hat{C}) \in \mathrm{supt}_4(\ell(e)) \\
\text{and } i \in Q(\ell(e)) \text{ and } u \in Q \text{ and } P \in \{=,\geq\} \text{ we have:} \\
\quad \begin{cases} \mathrm{inpo}(i,e^{**}((B,\tilde{B},\hat{B},\hat{C}))) \\ \mathrm{Pue}[B]\cdot e^{**}((B,\tilde{B},\hat{B},\hat{C},B,0)) \end{cases} \\
\quad \Leftrightarrow \begin{cases} \mathrm{inpo}(i,e^{**}((B,\tilde{B},\hat{B},\hat{C}))) + \mathrm{inpo}(ue(B),e^{**}((B,\tilde{B},\hat{B},\hat{C}))) \\ \mathrm{Pue}[B]e^{**}((B,\tilde{B},\hat{B},\hat{C},B,0)) + \mathrm{inpo}(ue(B),e^{**}((B,\tilde{B},\hat{B},\hat{C}))). \end{cases}
\end{cases}
$$

Since inpo is additive, we see that

$$
(73.1)\quad \begin{cases}
\text{for any } (B,\tilde{B},\hat{B},\hat{C}) \in \mathrm{supt}_4(\ell(e)) \\
\text{and any } i \in Q(\ell(e)) \text{ and } u \in Q \text{ we have:} \\
\mathrm{inpo}(i,e^{**}((B,\tilde{B},\hat{B},\hat{C}))) + \mathrm{inpo}(ue(B),e^{**}((B,\tilde{B},\hat{B},\hat{C}))) \\
= \mathrm{inpo}(i+ue(B),e^{**}((B,\tilde{B},\hat{B},\hat{C}))).
\end{cases}
$$

By (64) we see that

$$(73.2)\begin{cases}\text{if } (B,\widetilde{B},\hat{B},\hat{C}) \in \mathrm{supt}_4(\ell(e)) \text{ is such that} \\ e^{**}((B,\widetilde{B},\hat{B},\hat{C},B,0)) \neq 0 \\ \text{then for any } u \in Q \text{ we have:} \\ ue[B]e^{**}((B,\widetilde{B},\hat{B},\hat{C},B,0)) + \mathrm{inpo}(ue(B),e^{**}((B,\widetilde{B},\hat{B},\hat{C}))) \\ = u\ .\end{cases}$$

By (73.1) and (73.2) we see that

$$(73)\begin{cases}\text{if } (B,\widetilde{B},\hat{B},\hat{C}) \in \mathrm{supt}_4(\ell(e)) \text{ is such that} \\ e^{**}((B,\widetilde{B},\hat{B},\hat{C},B,0)) \neq 0 \\ \text{then for any } i \in Q(\ell(e)) \text{ and } u \in Q \text{ and } P \in \{=,\geq\} \\ \text{we have:} \\ \quad\begin{cases}\mathrm{inpo}(i,e^{**}((B,\widetilde{B},\hat{B},\hat{C}))) + \mathrm{inpo}(ue(B),e^{**}((B,\widetilde{B},\hat{B},\hat{C}))) \\ Pue[B]e^{**}((B,\widetilde{B},\hat{B},\hat{C},B,0)) + \mathrm{inpo}(ue(B),e^{**}((B,\widetilde{B},\hat{B},\hat{C})))\end{cases} \\ \quad\Leftrightarrow \begin{cases}\mathrm{inpo}(i+ue(B),e^{**}((B,\widetilde{B},\hat{B},\hat{C})) \\ Pu.\end{cases}\end{cases}$$

By (71), (72) and (73) we see that

$$(74)\begin{cases}\text{if } (B,\widetilde{B},\hat{B},\hat{C}) \in \mathrm{supt}_4(\ell(e)) \text{ is such that} \\ e^{**}((B,\widetilde{B},\hat{B},\hat{C},B,0)) \neq 0 \\ \text{then for any } i \in Q(\ell(e)) \text{ and } u \in Q \text{ and } P \in \{=,\geq\} \\ \text{we have:} \\ \quad\begin{cases}\mathrm{inpo}(i,e^{**}(B,\widetilde{B},\hat{B},\hat{C})) \\ Pue[B]\end{cases} \\ \quad\Leftrightarrow \begin{cases}\mathrm{inpo}(i+ue(B),e^{**}((B,\widetilde{B},\hat{B},\hat{C}))) \\ Pu\ .\end{cases}\end{cases}$$

By (48) and (74) we see that

$$(75)\quad \begin{cases} \text{if } (B,\widetilde{B},\hat{B},\hat{C}) \in \mathrm{supt}_4(\ell(e)) \text{ is such that} \\ (\widetilde{B},\hat{B},\hat{C}) \in \mathrm{supt}((e^{**},B)) \cap \mathrm{supt}(e^{**},B) \\ \text{then for any } i \in Q(\ell(e)) \text{ and } u \in Q \text{ and } P \in \{=,\geq\} \\ \text{we have:} \\ \mathrm{inpo}(i,e^{**}(B,\widetilde{B},\hat{B},\hat{C}))Pue[B] \\ \Leftrightarrow \mathrm{inpo}(i + ue(B),e^{**}((B,\widetilde{B},\hat{B},\hat{C})))Pu\ . \end{cases}$$

Since inpo is linear, by (63) we see that

$$(76)\quad \begin{cases} \text{if } (B,\widetilde{B},\hat{B},\hat{C}) \in \mathrm{supt}_4(\ell(e)) \text{ is such that} \\ (\widetilde{B},\hat{B},\hat{C}) \in \mathrm{supt}((e^{**},B))\setminus \mathrm{supt}(e^{**},B) \\ \text{then for any } i \in Q(\ell(e)) \text{ and } u \in Q \text{ we have:} \\ \mathrm{inpo}(i + ue(B),e^{**}((B,\widetilde{B},\hat{B},\hat{C}))) \\ = u + \mathrm{inpo}(i,e^{**}((B,\widetilde{B},\hat{B},\hat{C}))). \end{cases}$$

By (76) we see that

$$(77)\quad \begin{cases} \text{if } (B,\widetilde{B},\hat{B},\hat{C}) \in \mathrm{supt}_4(\ell(e)) \text{ is such that} \\ (\widetilde{B},\hat{B},\hat{C}) \in \mathrm{supt}((e^{**},B))\setminus \mathrm{supt}(e^{**},B) \\ \text{then for any } i \in Q(\ell(e)) \text{ and } u \in Q \text{ we have:} \\ \mathrm{inpo}(i + ue(B),e^{**}((B,\widetilde{B},\hat{B},\hat{C}))) \geq u. \end{cases}$$

By (62) we see that

$$(78)\quad \begin{cases} \text{if } e(B,\tilde{B},\hat{B},\hat{C}) \in \mathrm{supt}_4(\ell(e)) \text{ is such that} \\ (\tilde{B},\hat{B},\hat{C}) \in \mathrm{supt}(e^{**},B)\setminus \mathrm{supt}((e^{**},B)) \\ \text{then for any } i \in Q(\ell(e)) \text{ and } u \in Q \text{ we have:} \\ \mathrm{inpo}(i,e^{**}(B,\tilde{B},\hat{B},\hat{C})) \geq ue[B]. \end{cases}$$

By (34), (69), (70), (75), (76), (77) and (78) it follows that

$$(79)\quad \begin{cases} \text{for any } B \in [1,o(\ell(e))] \text{ and } u \in Q \text{ and } P \in \{=,\geq\} \\ \text{we have: } Q(e^{**}((B))Pu) = \{i + ue(B): i \in Q(e^{**}(B)Pue[B])\}. \end{cases}$$

By (3) and (79) we see that

$$(80)\quad \begin{cases} \text{for any } B \in [1,o(\ell(e))] \text{ and } B' \in [1,B] \text{ and } u \in Q \text{ and } P \in \{=,\geq\} \\ \text{we have: } Q(e^{**}((B))Pu) \cap Q(\ell(e),B') \\ \qquad = \{i + ue(B): i \in Q(e^{**}(B)Pue[B]) \cap Q(\ell(e),B')\}. \end{cases}$$

We shall now give a more detailed description of the supports of e^* and e^{**} when e is a parachip. Recall that

(81) if e is a parachip then $e[b] \neq 0$ for $1 \leq b \leq o(\ell(e))-2$

In view of (8), (9), (10) and (23), be decreasing induction on b we see that

$$(82)\begin{cases} \text{if } (B,\widetilde{B},\hat{B},\hat{C}) \in \operatorname{supt}_4(\ell(e)) \text{ is such that } \hat{C} \neq 0 < B \leq \widetilde{B} \leq \hat{B} \\ \text{and } e(\bar{B},\hat{B},\hat{C}) = 0 \text{ for all } \bar{B} \in [B+1,\widetilde{B}] \\ \text{then: for any } (b,c) \in \operatorname{supt}_2(\ell(e)) \text{ we have} \\ e^*(B,\widetilde{B},\hat{B},\hat{C},b,c) = \begin{cases} 0 & \text{if } (b,c) \neq (\hat{B},\hat{C}) \\ 1 & \text{if } (b,c) = (\hat{B},\hat{C}) \end{cases} \\ \text{and } e^*((B,\widetilde{B},\hat{B},\hat{C},b,c)) = \begin{cases} 0 & \text{if } b < B \\ e(b,\hat{B},\hat{C}) & \text{if } b \geq B. \end{cases} \end{cases}$$

By (31), (32) and (82) we see that

$$(83)\begin{cases} \text{if } (B,\widetilde{B},\hat{B},\hat{C}) \in \operatorname{supt}_4(\ell(e)) \text{ is such that } \hat{C} \neq 0 < B \leq \widetilde{B} \leq \hat{B} \\ \text{and } e(\bar{B},\hat{B},\hat{C}) = 0 \text{ for all } \bar{B} \in [B+1,\widetilde{B}]. \\ \text{then: } (\widetilde{B},\hat{B},\hat{C}) \in \operatorname{supt}(e^*,B)\setminus\operatorname{supt}(e^{**},B) \\ \text{whereas:} \\ (\widetilde{B},\hat{B},\hat{C}) \in \operatorname{supt}((e^*,B)) \Leftrightarrow e(b,\hat{B},\hat{C}) \neq 0 \text{ for some } b \geq B \\ \text{and moreover:} \\ (\widetilde{B},\hat{B},\hat{C}) \in \operatorname{supt}((e^{**},B)) \\ \quad \Leftrightarrow e(B,\hat{B},\hat{C}) \neq 0 \\ \quad \Leftrightarrow \begin{cases} e(B,\hat{B},\hat{C}) \neq 0 \text{ and } e^{**}((B,\widetilde{B},\hat{B},\hat{C},\hat{B},\hat{C})) = e(B,\hat{B},\hat{C})^{-1} \\ \text{and} \begin{cases} e^{**}((B,\widetilde{B},\hat{B},\hat{C},b,c)) = 0 \\ \text{for all } (b,c) \in \operatorname{supt}_2(\ell(e))\setminus\{(\hat{B},\hat{C})\}. \end{cases} \end{cases} \end{cases}$$

In view of (2), (8), (9), (10) and (23), by decreasing induction on b we see that

$$(84)\begin{cases} \text{if } (B,\tilde{B},\hat{B},\hat{C}) \in \text{supt}_4(\ell(e)) \text{ is such that } \hat{C} \neq 0 < B \leq \tilde{B} \leq \hat{B} \\ \text{and } e(\bar{B},\hat{B},\hat{C}) \neq 0 \text{ for some } \bar{B} \in [B+1,\tilde{B}] \\ \text{then: upon letting } B^* = \max\{\bar{B} \in [B+1,\tilde{B}]: e(\bar{B},\hat{B},\hat{C}) \neq 0\} \\ \text{we have } B < B^* \leq \min(\tilde{B},o(\ell(e))-1) \\ \text{and for any } (b,c) \in \text{supt}_2(\ell(e)) \text{ we have} \\ e^*(B,\tilde{B},\hat{B},\hat{C},b,c) = \begin{cases} 0 \text{ if } b < B \\ 0 \text{ if } b \geq B^* \text{ and } (b,c) \neq (\hat{B},\hat{C}) \\ 1 \text{ if } (b,c) = (\hat{B},\hat{C}) \\ e(B^*,\hat{B},\hat{C}) \neq 0 \text{ if } b = B^*-1 \end{cases} \\ \text{and } e^*((B,\tilde{B},\hat{B},\hat{C},b,c)) = \begin{cases} 0 \text{ if } b < B \\ e(b,\hat{B},\hat{C}) \text{ if } b \geq B^* \\ e(B^*,\hat{B},\hat{C}) \neq 0 \text{ if } b = B^* . \end{cases} \end{cases}$$

In view of (21), (22), (81) and (84), by decreasing induction on b we see that

$$(85)\begin{cases} \text{if } e \text{ is a parachip and} \\ \text{if } (B,\tilde{B},\hat{B},\hat{C}) \in \text{supt}_4(\ell(e)) \text{ is such that } \hat{C} \neq 0 < B \leq \tilde{B} \leq \hat{B} \\ \text{and } e(\bar{B},\hat{B},\hat{C}) \neq 0 \text{ for some } \bar{B} \in [B+1,\tilde{B}] \\ \text{then: upon letting } B^* = \max\{\bar{B} \in [B+1,\tilde{B}]: e(\bar{B},\hat{B},\hat{C}) \neq 0\} \\ \text{we have } B < B^* \leq \min(\tilde{B},o(\ell(e))-1) \\ \text{and for all } (b,c) \in \text{supt}_2(\ell(e)) \text{ with } B \leq b < B^* \\ \text{we have } e^*(B,\tilde{B},\hat{B},\hat{C},b,c) \neq 0 \neq e^*((B,\tilde{B},\hat{B},\hat{C},b,c)). \end{cases}$$

By (31), (32), (84) and (85) we see that

$$
(86)\quad \begin{cases}
\text{if } e \text{ is a parachip and} \\
\text{if } (B,\tilde{B},\hat{B},\hat{C}) \in \mathrm{supt}_4(\ell(e)) \text{ is such that } \hat{C} \neq 0 < B \le \tilde{B} \le \hat{B} \\
\text{and } e(\bar{B},\hat{B},\hat{C}) \neq 0 \text{ for some } \bar{B} \in [B+1,\tilde{B}] \\
\text{then:} \\
(\tilde{B},\hat{B},\hat{C}) \in \mathrm{supt}(e^*,B) \cap \mathrm{supt}((e^*,B)) \cap \mathrm{supt}(e^{**},B) \cap \mathrm{supt}((e^{**},B)) \\
\text{and upon letting } B^* = \max\{\bar{B} \in [B+1,\tilde{B}] : e(\bar{B},\hat{B},\hat{C}) \neq 0\} \\
\text{we have } B < B^* \le \tilde{B} \text{ and we have} \\
\{(b,c) \in \mathrm{supt}_2(\ell(e)) : e^*(B,\tilde{B},\hat{B},\hat{C},b,c) \neq 0\} \\
= \{(b,c) \in \mathrm{supt}_2(\ell(e)) : e^{**}(B,\tilde{B},\hat{B},\hat{C},b,c) \neq 0\} \\
= \{(b,c) \in \mathrm{supt}_2(\ell(e)) : e^{**}((B,\tilde{B},\hat{B},\hat{C},b,c)) \neq 0\} \\
= \{(b,c) \in \mathrm{supt}_2(\ell(e)) : \text{either } B \le b < B^* \text{ or } (b,c) = (\hat{B},\hat{C})\} \\
\subset \{(b,c) \in \mathrm{supt}_2(\ell(e)) : \text{either } e^*((B,\tilde{B},\hat{B},\hat{C},b,c)) \neq 0 \\
\qquad\qquad \text{or } (b,c) = (\hat{B},\hat{C})\}.
\end{cases}
$$

By (8), (26), (33) and (34) we see that

$$
(91)\quad \begin{cases}
\text{if } (B,\tilde{B},\hat{B},\hat{C}) \in \mathrm{supt}_4(\ell(e)) \text{ is such that } \hat{C} = 0 < B \le \tilde{B} = \hat{B} = o(\ell(e)) \\
\text{then:} \\
(\tilde{B},\hat{B},\hat{C}) \in \mathrm{supt}(e^*,B) \cap \mathrm{supt}((e^*,B)) \cap \mathrm{supt}(e^{**},B) \cap \mathrm{supt}((e^{**},B)) \\
\text{and:} \\
\{(b,c) \in \mathrm{supt}_2(\ell(e)) : e^*(B,\tilde{B},\hat{B},\hat{C},b,c) \neq 0\} \\
= \{(b,c) \in \mathrm{supt}_2(\ell(e)) : e^{**}(B,\tilde{B},\hat{B},\hat{C},b,c) \neq 0\} \\
= \{(b,c) \in \mathrm{supt}_2(\ell(e)) : e^{**}((B,\tilde{B},\hat{B},\hat{C},b,c)) \neq 0\} \\
= \{(b,c) \in \mathrm{supt}_2(\ell(e)) : e^*((B,\tilde{B},\hat{B},\hat{C},b,c)) \neq 0\} \\
= \{(b,c) \in \mathrm{supt}_2(\ell(e)) : b \ge B\}.
\end{cases}
$$

In view of (8), (9), (10) and (24), by decreasing induction on b we see that

$$(92)\begin{cases} \text{if } (B,\widetilde{B},\hat{B},\hat{C}) \in \mathrm{supt}_4(\ell(e)) \text{ is such that } \hat{C}=0<\widetilde{B}\le B=\hat{B}=o(\ell(e))-1 \\ \text{and } e(\bar{B},\widetilde{B},\widetilde{C}) = 0 \text{ for all } \bar{B}\in[B+1,\widetilde{B}] \text{ and all } \widetilde{C}\in[0,\widetilde{B}(\ell(e))] \\ \text{then: for all } (b,c)\in\mathrm{supt}_2(\ell(e)) \text{ we have} \\ e^*(B,\widetilde{B},\hat{B},\hat{C},b,c) = \begin{cases} 0 & \text{if } b\ne\hat{B} \\ 1 & \text{if } b=\hat{B} \end{cases} \\ \text{and } e^*((B,\widetilde{B},\hat{B},\hat{C},b,c)) = \begin{cases} 0 & \text{if } b\ne B \\ \sum\limits_{0\le C'\le\widetilde{B}(\ell(e))} e(B,\widetilde{B},C') & \text{if } b=B \end{cases} \end{cases}$$

By (31), (32) and (92) we see that

$$(93)\begin{cases} \text{if } (B,\widetilde{B},\hat{B},\hat{C}) \in \mathrm{supt}_4(\ell(e)) \text{ is such that } \hat{C}=0<B\le\widetilde{B}=\hat{B}=o(\ell(e))-1 \\ \text{and } e(\bar{B},\widetilde{B},\widetilde{C}) = 0 \text{ for all } \bar{B}\in[B+1,\widetilde{B}] \text{ and all } \widetilde{C}\in[0,\widetilde{B}(\ell(e))] \\ \text{then: } (\widetilde{B},\hat{B},\hat{C})\in\mathrm{supt}(e^*,B) \\ \text{whereas:} \\ (\widetilde{B},\hat{B},\hat{C})\in\mathrm{supt}(e^{**},B) \\ \Leftrightarrow B=\widetilde{B} \\ \Leftrightarrow \begin{cases} B=\widetilde{B} \text{ and for all } (b,c)\in\mathrm{supt}_2(\ell(e)) \text{ we have} \\ e^{**}(B,\widetilde{B},\hat{B},\hat{C},b,c) = \begin{cases} 0 & \text{if } b\ne B \\ 1 & \text{if } b=B \end{cases} \end{cases} \\ \text{and upon letting } V = \sum\limits_{0\le C'\le\widetilde{B}(\ell(e))} e(B,\widetilde{B},C') \text{ we have:} \\ (\widetilde{B},\hat{B},\hat{C})\in\mathrm{supt}((e^*,B)) \\ \Leftrightarrow (\widetilde{B},\hat{B},\hat{C})\in\mathrm{supt}((e^{**},B)) \\ \Leftrightarrow e(B,\widetilde{B},C^*)\ne 0 \text{ for some } C^*\in[0,\widetilde{B}(\ell(e))] \\ \Leftrightarrow V\ne 0 \\ \Leftrightarrow \begin{cases} V\ne 0 \text{ and for all } (b,c)\in\mathrm{supt}_2(\ell(e)) \text{ we have} \\ e^{**}((B,\widetilde{B},\hat{B},\hat{C},b,c)) = \begin{cases} 0 & \text{if } b\ne\hat{B} \\ V^{-1} & \text{if } b=\hat{B}\ . \end{cases} \end{cases} \end{cases}$$

In view of (8), (9), (10) and (24), by decreasing induction on b we see that

$$
(94)\quad
\begin{cases}
\text{if } (B,\tilde{B},\hat{B},\hat{C}) \in \mathrm{supt}_4(\ell(e)) \text{ is such that } \hat{C} = 0 < B \le \tilde{B} = \hat{B} = o(\ell((e))-1 \\
\text{and } e(\bar{B},\tilde{B},\tilde{C}) \ne 0 \text{ for some } \bar{B} \in [B+1,\tilde{B}] \text{ and some } \tilde{C} \in [0,\tilde{B}(\ell(e))] \\
\text{then: upon letting } B^* = \max\{\bar{B} \in [B+1,\tilde{B}] : e(\bar{B},\tilde{B},\tilde{C}) \ne 0 \\
\qquad\qquad \text{for some } \tilde{C} \in [0,\tilde{B}(\ell(e))]\} \\
\text{we have } B < B^* \le \tilde{B} \text{ and for any } (b,c) \in \mathrm{supt}_2(\ell(e)) \text{ we have} \\
e^*(B,\tilde{B},\hat{B},\hat{C},b,c) = \begin{cases} 0 \text{ if } b < B \\ 0 \text{ if } B^* \le b \ne \tilde{B} \\ 1 \text{ if } b = \tilde{B} \\ \sum_{0 \le C' \le \tilde{B}(\ell(e))} e(B^*,\tilde{B},C') \ne 0 \text{ if } b = B^*-1 \end{cases} \\
\text{and } e^*((B,\tilde{B},\hat{B},\hat{C},b,c)) = \begin{cases} 0 \text{ if } b < B \\ 0 \text{ if } b > B^* \\ \sum_{0 \le C' \le \tilde{B}(\ell(e))} e(B^*,\tilde{B},C') \ne 0 \text{ if } b = B^*. \end{cases}
\end{cases}
$$

In view of (21), (22), (81) and (94), by decreasing induction on b we see that

$$
(95)\quad
\begin{cases}
\text{if } e \text{ is a parachip and} \\
\text{if } (B,\tilde{B},\hat{B},\hat{C}) \in \mathrm{supt}_4(\ell(e)) \text{ is such that } \hat{C} = 0 < B \le \tilde{B} = \hat{B} = o(\ell(e))-1 \\
\text{and } e(\bar{B},\tilde{B},\tilde{C}) \ne 0 \text{ for some } \bar{B} \in [B+1,\tilde{B}] \text{ and some } \tilde{C} \in [0,\tilde{B}(\ell(e))] \\
\text{then upon letting } B^* = \max\{\bar{B} \in [B+1,\tilde{B}] : e(\bar{B},\tilde{B},C) \ne 0 \\
\qquad\qquad \text{for some } \tilde{C} \in [0,\tilde{B}(\ell(e))]\} \\
\text{we have } B < B^* \le \tilde{B} \\
\text{and for all } (b,c) \in \mathrm{supt}_2(\ell(e)) \text{ with } B \le b < B^* \text{ we have} \\
e^*(B,\tilde{B},\hat{B},\hat{C},b,c) \ne 0 \ne e^*((B,\tilde{B},\hat{B},\hat{C},b,c)).
\end{cases}
$$

By (31), (32), (94) and (95) we see that

$$(96)\quad \begin{cases} \text{if } e \text{ is a parachip and} \\ \text{if } (B,\tilde{B},\hat{B},\hat{C}) \in \text{supt}_4(\ell(e)) \text{ is such that } \hat{C} = 0 < B \le \tilde{B} = \hat{B} = o(\ell(e))-1 \\ \text{and } e(\bar{B},\tilde{B},\tilde{C}) \ne 0 \text{ for some } \bar{B} \in [B+1,B] \text{ and some } \tilde{C} \in [0,\tilde{B}(\ell(e))] \\ \text{then:} \\ (\tilde{B},\hat{B},\hat{C}) \in \text{supt}(e^*,B) \cap \text{supt}((e^*,B)) \cap \text{supt}(e^{**},B) \cap \text{supt}((e^{**},B)) \\ \text{and upon letting } B^* = \max\{\bar{B} \in [B+1,\tilde{B}]: e(\bar{B},\tilde{B},\tilde{C}) \ne 0 \\ \qquad\qquad \text{for some } \tilde{C} \in [0,\tilde{B}(\ell(e))]\} \\ \text{we have:} \\ \{(b,c) \in \text{supt}_2(\ell(e)): e^*(B,\tilde{B},\hat{B},\hat{C},b,c) \ne 0\} \\ = \{(b,c) \in \text{supt}_2(\ell(e)): e^{**}(B,\tilde{B},\hat{B},\hat{C},b,c) \ne 0\} \\ = \{(b,c) \in \text{supt}_2(\ell(e)): e^{**}((B,\tilde{B},\hat{B},\hat{C},b,c)) \ne 0\} \\ = \{(b,c) \in \text{supt}_2(\ell(e)): \text{either } b \in [B,B^*-1] \text{ or } b = \tilde{B}\} \\ \subset \{(b,c) \in \text{supt}_2(\ell(e)): \text{either } e^*((B,\tilde{B},\hat{B},\hat{C},b,c)) \ne 0 \text{ or } b = \tilde{B}\}. \end{cases}$$

The description of the supports of e^* and e^{**} given in (82) to (86) and (91) to (96) is now completed by noting that in view of (12) and (42) we have

$$(97)\quad \begin{cases} \text{supt}(e^*) \cup \text{supt}((e^*)) \cup \text{supt}(e^{**}) \cup \text{supt}((e^{**})) \\ \subset \{(\tilde{B},\hat{B},\hat{C}) \in \text{supt}_3(\ell(e)): \text{either } \hat{C} \ne 0 < B \le \tilde{B} \le \hat{B} \\ \qquad\qquad \text{or } \hat{C} = 0 < B \le \tilde{B} = \hat{B} \ge o(\ell(e))-1\}. \end{cases}$$

§40. Isobars for derived scales

Let e be a protochip, let R be a ring, let Y be an indeterminate net over R with $\ell(Y) = \ell(e)$, and let $u \in Q$.

By (35) of §39 we know that

(11) e^{**} is a scale

and so the definitions and observations made in §38 are applicable. By (50) of §39 we see that

$$(12)\quad \begin{cases} \text{if } 1 \le B = o(\ell(e))-1 \text{ and } B' \in [1,B] \text{ and } P \in \{=,\ge\} \text{ then} \\ Y\langle B',B\rangle^{u}_{(R,e^{**}P)Q} = Y\langle B',B\rangle^{u}_{(RP)Q} . \end{cases}$$

By (80) of §39 we see that

$$(13)\quad \begin{cases} \text{if } B \in [1,o(\ell(e))-1] \text{ and } P \in \{=,\ge\} \text{ then} \\ Y\langle B,B\rangle^{u}_{((R,e^{**}P))Q} = Y^{ue(B)}Y\langle B,B\rangle^{ue[B]}_{(R,e^{**}P)Q} \end{cases}$$

and by (55) of §39 we see that

$$(14)\quad \begin{cases} \text{if } B \in [1,o(\ell(e))-2] \text{ and } P \in \{=,\ge\} \text{ then} \\ Y\langle B,B\rangle^{u}_{(R,e^{**}P)Q} \\ = \sum\limits_{\substack{v\in Q \text{ and } w\in Q \\ \text{with } v+w=u}} Y\langle B,B\rangle^{v}_{(RP)Q}Y\langle B+1,B+1\rangle^{w}_{((R,e^{**}P))Q} . \end{cases}$$

In view of (3) and (5) of §38, by (13) we see that

$$(15)\quad \begin{cases} \text{if } B \in [1,O(\ell(e))-1] \text{ and } B' \in [1,B] \text{ and } P \in \{=,\geq\} \text{ then} \\ Y\langle B',B\rangle^{u}_{((R,e{*}{*}P))Q} = Y^{ue(B)}Y\langle B',B\rangle^{ue[B]}_{(R,e{*}{*}P)Q} \end{cases}$$

and by (14) we see that

$$(16)\quad \begin{cases} \text{if } B \in [1,o(\ell(e))-2] \text{ and } B' \in [1,B] \text{ and } P \in \{=,\geq\} \text{ then} \\ Y\langle B',B\rangle^{u}_{(R,e{*}{*}P)Q} \\ = \sum\limits_{\substack{v\in Q \text{ and } w\in Q \\ \text{with } v+w=u}} Y\langle B',B\rangle^{v}_{(RP)Q} Y\langle B',B+1\rangle^{w}_{((R,e{*}{*}P))Q} \,. \end{cases}$$

In view of (12), (15) and (16), by decreasing induction on B we see that

$$(17)\quad \begin{cases} \text{if } B \in [1,o(\ell(e))-1] \text{ and } B' \in [1,B] \text{ and } P \in \{=,\geq\} \text{ then} \\ Y\langle B',B\rangle^{u}_{(R,eP)Q} = Y\langle B',B\rangle^{u}_{(R,e{*}{*}P)Q} \\ \text{and} \\ Y\langle B',B\rangle^{u}_{((R,eP))Q} = Y\langle B',B\rangle^{u}_{((R,e{*}{*}P))Q} \,. \end{cases}$$

By (17) we also see that

$$(18)\quad \begin{cases} \text{if } B \in [1,o(\ell(e))-1] \text{ and } B' \in [1,B] \text{ and } P \in \{=,\geq\} \text{ then} \\ Y\langle B',B\rangle^{u}_{(R,eP)} = Y\langle B',B\rangle^{u}_{(R,e{*}{*}P)} \\ \text{and} \\ Y\langle B',B\rangle^{u}_{((R,eP))} = Y\langle B',B\rangle^{u}_{((R,e{*}{*}P))} \,. \end{cases}$$

§41. Isobars and initial forms for scales

Let E be a scale. Let R be a ring and let Y be an indeterminate net over R with $\ell(Y) = \ell(E)$. Let $B \in [1,o(\ell(E))-1]$ and $B' \in [1,B]$. Let $u \in Q$. Let $P \in \{=,\geq,>\}$.

Given any ring-homomorphism $g: R \to R'$ we define

$$\mathrm{Sub}[g,Y\langle B',B\rangle,EPu]_Q: Y\langle B',B\rangle^u_{(R,EP)Q} \to R'[Y]_Q$$

$$\mathrm{Sub}[[g,Y\langle B',B\rangle,EPu]]_Q: Y\langle B',B\rangle^u_{((R,EP))Q} \to R'[Y]_Q$$

$$\mathrm{Sub}[g,Y\langle B',B\rangle,EPu]: Y\langle B',B\rangle^u_{(R,EP)} \to R'[Y]$$

and

$$\mathrm{Sub}[[g,Y\langle B',B\rangle,EPu]]: Y\langle B',B\rangle^u_{((R,EP))} \to R'[Y]$$

to be the g-homomorphisms induced by $\mathrm{Sub}[g,Y]_Q$ and we observe that

$$\mathrm{Sub}[g,Y]_Q(Y\langle B',B\rangle^u_{(R,EP)Q}) = Y\langle B',B\rangle^u_{(g(R),EP)Q}$$

$$\mathrm{Sub}[g,Y]_Q(Y\langle B',B\rangle^u_{((R,EP))Q}) = Y\langle B',B\rangle^u_{((g(R),EP))Q}$$

$$\mathrm{Sub}[g,Y]_Q(Y\langle B',B\rangle^u_{(R,EP)}) = Y\langle B',B\rangle^u_{(g(R),EP)}$$

and

$$\mathrm{Sub}[g,Y]_Q(Y\langle B',B\rangle^u_{((R,EP))}) = Y\langle B',B\rangle^u_{((g(R),EP))}$$

and we define

$$\mathrm{Sub}[g,Y\langle B',B\rangle,EPu]^{*}_{Q}\colon Y\langle B',B\rangle^{u}_{(R,EP)Q} \to Y\langle B',B\rangle^{u}_{(g(R),EP)Q}$$

$$\mathrm{Sub}[[g,Y\langle B',B\rangle,EPu]]^{*}_{Q}\colon Y\langle B',B\rangle^{u}_{((R,EP))Q} \to Y\langle B',B\rangle^{u}_{((g(R),EP))Q}$$

$$\mathrm{Sub}[g,Y\langle B',B\rangle,EPu]^{*}\colon Y\langle B',B\rangle^{u}_{(R,EP)} \to Y\langle B',B\rangle^{u}_{(g(R),EP)}$$

and

$$\mathrm{Sub}[[g,Y\langle B',B\rangle,EPu]]^{*}\colon Y\langle B',B\rangle^{u}_{((R,EP))} \to Y\langle B',B\rangle^{u}_{((g(R),EP))}$$

to be the g-epimorphisms induced by $\mathrm{Sub}[g,Y]_Q$.

We define the R-homomorphism

$$\mathrm{Iso}[R,Y\langle B',B\rangle,EPu]_Q\colon R[Y\langle B'\rangle]_Q \to R[Y]_Q$$

by putting, for all $F \in R[Y\langle B'\rangle]$,

$$\mathrm{Iso}[R,Y\langle B',B\rangle,EPu]_Q(F)$$

$$= \begin{cases} \sum\limits_{j\in Q(E(B)Pu)\cap Q(\ell(E),B')} F[j]Y^{j} & \text{if } P \in \{\geq\} \\ \sum\limits_{j\in Q(E(B)Pu)\cap Q(\ell(E),B)} F[j]Y^{j} & \text{if } P \in \{=,>\} \end{cases}$$

and we define

$$\mathrm{Iso}[R,Y\langle B',B\rangle,EPu]\colon R[Y\langle B'\rangle] \to R[Y]$$

to be the R-homomorphism induced by $\mathrm{Iso}[R,Y\langle B',B\rangle,EPu]_Q$ and we observe that

$$\mathrm{Iso}[R,Y\langle B',B\rangle,EPu]_Q(R[Y\langle B'\rangle])_Q = Y\langle B',B\rangle^{u}_{(R,EP)Q}$$

and

$$\text{Iso}[R,Y\langle B',B\rangle,EPu]_Q(R[Y\langle B'\rangle]) = Y\langle B',B\rangle^u_{(R,EP)}$$

and we define

$$\text{Iso}[R,Y\langle B',B\rangle,EPu]^*_Q: R[Y\langle B'\rangle]_Q \to Y\langle B',B\rangle^u_{(R,EP)Q}$$

and

$$\text{Iso}[R,Y\langle B',B\rangle,EPu]^*: R[Y\langle B'\rangle] \to Y\langle B',B\rangle^u_{(R,EP)}$$

to be the R-epimorphisms induced by $\text{Iso}[R,Y\langle B',B\rangle,EPu]_Q$.

We define the R-homomorphism

$$\text{Iso}[[R,Y\langle B',B\rangle,EPu]]_Q: R[Y\langle B'\rangle]_Q \to R[Y]_Q$$

by putting, for all $F \in R[Y\langle B'\rangle]_Q$,

$$\text{Iso}[[R,Y\langle B',B\rangle,EPu]]_Q(F)$$

$$= \begin{cases} \displaystyle\sum_{j\in Q(E((B))Pu)\cap Q(\ell(E),B')} F[j]Y^j & \text{if } P \in \{\geq\} \\ \displaystyle\sum_{j\in Q(E((B))Pu)\cap Q(\ell(E),B)} F[j]Y^j & \text{if } P \in \{=,>\} \end{cases}$$

and we define

$$\text{Iso}[[R,Y\langle B',B\rangle,EPu]]: R[Y\langle B'\rangle] \to R[Y]$$

to be the R-homomorphism induced by $\text{Iso}[[R,Y\langle B',B\rangle,EPu]]_Q$ and we observe that

$$\mathrm{Iso}[[R,Y\langle B',B\rangle,EPu]]_Q(R[Y\langle B'\rangle]_Q) = Y\langle B',B\rangle^u_{((R,EP))Q}$$

and

$$\mathrm{Iso}[[R,Y\langle B',B\rangle,EPu]]_Q(R[Y\langle B'\rangle]) = Y\langle B',B\rangle^u_{((R,EP))}$$

and we define

$$\mathrm{Iso}[[R,Y\langle B',B\rangle,EPu]]^*_Q\colon R[Y\langle B'\rangle]_Q \to Y\langle B',B\rangle^u_{((R,EP))Q}$$

and

$$\mathrm{Iso}[[R,Y\langle B',B\rangle,EPu]]^*\colon R[Y\langle B'\rangle] \to Y\langle B',B\rangle_{((R,EP))}$$

to be the R-epimorphisms induced by $\mathrm{Iso}[[R,Y\langle B',B\rangle,EPu]]_Q$.

We define

$$\mathrm{Info}[R,Y\langle B,B\rangle,E=u]_Q\colon Y\langle B,B\rangle^u_{(R,E\geq)Q} \to R[Y]_Q$$

and

$$\mathrm{Info}[R,Y\langle B,B\rangle,E=u]\colon Y\langle B,B\rangle^u_{(R,E\geq)} \to R[Y]$$

to be the R-homomorphisms induced by $\mathrm{Iso}[R,Y\langle B,B\rangle,E=u]_Q$ and we note that they coincide with the R-homomorphisms induced by $\mathrm{Info}[R,Y,E(B,o(\ell(E)),o(\ell(E)),0)=u]_Q$ and we observe that

$$\mathrm{Info}[R,Y\langle B,B\rangle,E=u]_Q(Y\langle B,B\rangle^u_{(R,E\geq)Q}) = Y\langle B,B\rangle^u_{(R,E=)Q}$$

and

$$\mathrm{Info}[R,Y\langle B,B\rangle,E=u]_Q(Y\langle B,B\rangle_{(R,E\geq)}) = Y\langle B,B\rangle^u_{(R,E=)}$$

and we define

$$\text{Info}[R,Y\langle B,B\rangle,E=u]^*_Q\colon Y\langle B,B\rangle^u_{(R,E\geq)Q} \to Y\langle B,B\rangle^u_{(R,E=)Q}$$

and

$$\text{Info}[R,Y\langle B,B\rangle,E=u]^*\colon Y\langle B,B\rangle^u_{(R,E\geq)} \to Y\langle B,B\rangle^u_{(R,E=)}$$

to be the R-epimorphisms induced by $\text{Iso}[R,Y\langle B,B\rangle,E=u]_Q$ and we observe that

$$\ker(\text{Info}[R,Y\langle B,B\rangle,E=u]_Q) = \ker(\text{Info}[R,Y\langle B,B\rangle,E=u]^*_Q)$$
$$= Y\langle B,B\rangle^u_{(R,E>)Q}$$

and

$$\ker(\text{Info}[R,Y\langle B,B\rangle,E=u]) = \ker(\text{Info}[R,Y\langle B,B\rangle,E=u]^*)$$
$$= Y\langle B,B\rangle^u_{(R,E>)}.$$

We define

$$\text{Info}[[R,Y\langle B,B\rangle,E=u]]_Q\colon Y\langle B,B\rangle^u_{((R,E\geq))Q} \to R[Y]_Q$$

and

$$\text{Info}[[R,Y\langle B,B\rangle,E=u]]\colon Y\langle B,B\rangle^u_{((R,E\geq))} \to R[Y]$$

to be the R-homomorphisms induced by $\text{Iso}[[R,Y\langle B,B\rangle,E=u]]_Q$ and we note that they coincide with the R-homomorphisms induced by $\text{Info}[R,Y,E((B,o(\ell(E)),o(\ell(E)),0))=u]_Q$ and we observe that

$$\text{Info}[[R,Y\langle B,B\rangle,E=u]]_Q(Y\langle B,B\rangle^u_{((R,E\geq))Q}) = Y\langle B,B\rangle^u_{((R,E=))Q}$$

and

$$\text{Info}[[R,Y\langle B,B\rangle,E=u]]_Q(Y\langle B,B\rangle^u_{((R,E\geq))}) = Y\langle B,B\rangle^u_{((R,E=))}$$

and we define

$$\text{Info}[[R,Y\langle B,B\rangle,E=u]]^*_Q\colon Y\langle B,B\rangle^u_{((R,E\geq))Q} \to Y\langle B,B\rangle^u_{((R,E=))Q}$$

and

$$\text{Info}[[R,Y\langle B,B\rangle,E=u]]^*\colon Y\langle B,B\rangle^u_{((R,E\geq))} \to Y\langle B,B\rangle^u_{((R,E=))}$$

to be the R-epimorphisms induced by $\text{Iso}[[R,Y\langle B,B\rangle,E=u]]_Q$ and we observe that

$$\ker(\text{Info}[[R,Y\langle B,B\rangle,E=u]]_Q) = \ker(\text{Info}[[R,Y\langle B,B\rangle,E=u]]^*_Q)$$

$$= Y\langle B,B\rangle^u_{((R,E>))Q}$$

and

$$\ker(\text{Info}[[R,Y\langle B,B\rangle,E=u]]) = \ker(\text{Info}[[R,Y\langle B,B\rangle,E=u]]^*)$$

$$= Y\langle B,B\rangle^u_{((R,E>))}\ .$$

Given any ring-homomorphisms $g\colon R \to R'$ we define

$$\text{Info}[g,Y\langle B,B\rangle,E=u]_Q\colon Y\langle B,B\rangle^u_{(R,E\geq)Q} \to R'[Y]_Q$$

and

$$\text{Info}[g,Y\langle B,B\rangle,E=u]\colon Y\langle B,B\rangle^u_{(R,E\geq)} \to R'[Y]$$

to be the g-homomorphisms induced by
$Info[g,Y,E(B,o(\ell(E)),o(\ell(E)),0) = u]_Q$ and we observe that their images are

$$Y\langle B,B\rangle^u_{(g(R),E=)Q} \quad \text{and} \quad Y\langle B,B\rangle^u_{(g(R),E=)}$$

respectively and we define

$$Info[g,Y\langle B,B\rangle,E = u]^*_Q\colon Y\langle B,B\rangle^u_{(R,E\geq)Q} \to Y\langle B,B\rangle^u_{(g(R),E=)Q}$$

and

$$Info[g,Y\langle B,B\rangle,E = u]^*\colon Y\langle B,B\rangle^u_{(R,E\geq)} \to Y\langle B,B\rangle^u_{(g(R),E=)}$$

to be the g-epimorphisms induced by
$Info[g,Y,E(B,o(\ell(E)),o(\ell(E)),0) = u]_Q$ and we also define

$$Info[[g,Y\langle B,B\rangle,E = u]]_Q\colon Y\langle B,B\rangle^u_{((R,E\geq))Q} \to R'[Y]_Q$$

and

$$Info[[g,Y\langle B,B\rangle,E = u]]\colon Y\langle B,B\rangle^u_{((R,E\geq))} \to R'[Y]$$

to be the g-homomorphisms induced by
$Info[g,Y,E((B,o(\ell(E)),o(\ell(E)),0)) = u]_Q$ and we observe that their images are

$$Y\langle B,B\rangle^u_{((g(R),E=))Q} \quad \text{and} \quad Y\langle B,B\rangle^u_{((g(R),E=))}$$

respectively and we define

$$\text{Info}[[g,Y\langle B,B\rangle,E = u]]^{*}_{Q}\colon \quad Y\langle B,B\rangle^{u}_{((R,E\geq))Q} \to Y\langle B,B\rangle^{u}_{((g(R),E=))Q}$$

and

$$\text{Info}[[g,Y\langle B,B\rangle,E = u]]^{*}\colon \quad Y\langle B,B\rangle^{u}_{((R,E\geq))} \to Y\langle B,B\rangle^{u}_{((g(R),E=))}$$

to be the g-epimorphisms induced by $\text{Info}[g,Y,E((B,o(\ell(E)),o(\ell(E)),0)) = u]_{Q}$ and we note that we have the following four commutative diagrams whereby the first is

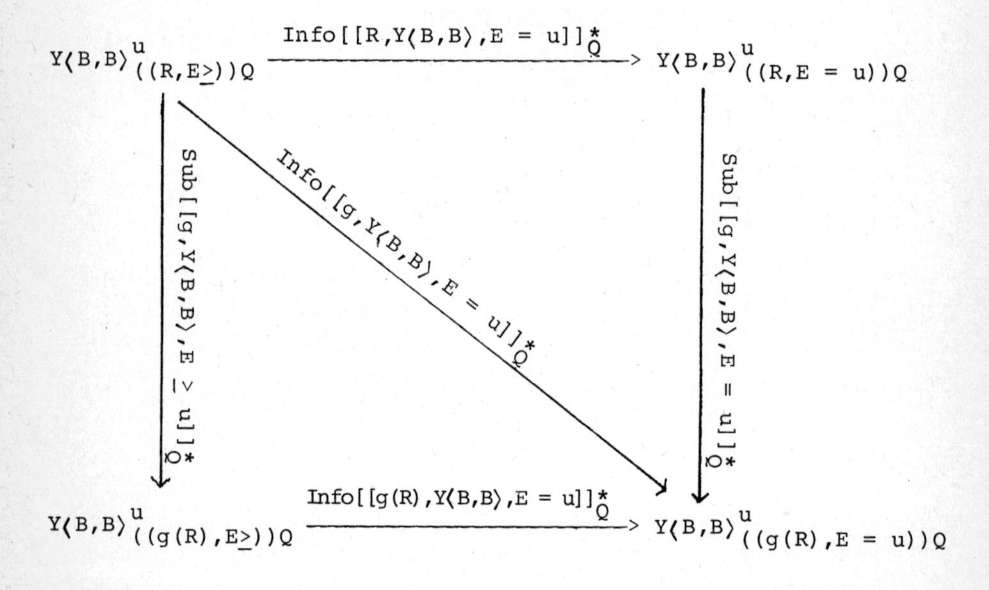

whereas the second is obtained from the first by everywhere replacing [[ ]] and (( )) by [ ] and ( ) respectively while the third is obtained from the first by everywhere deleting Q and finally the fourth is obtained from the second by everywhere deleting Q.

Given any R-net y with $\ell(y) = \ell(E)$ we define

$$\mathrm{sub}[R,Y\langle B',B\rangle = y, EPu]: Y\langle B',B\rangle^{u}_{(R,EP)} \to R$$

and

$$\mathrm{sub}[[R,Y\langle B',B\rangle = y, EPu]]: Y\langle B',B\rangle^{u}_{((R,EP))} \to R$$

to be the R-homomorphisms induced by $\mathrm{sub}[R,Y=y]$ and we put

$$\mathrm{iso}(R,y\langle B',B\rangle,EPu) = \mathrm{sub}[R,Y=y](Y\langle B',B\rangle^{u}_{(R,EP)})$$

and

$$\mathrm{iso}((R,y\langle B',B\rangle,EPu)) = \mathrm{sub}[R,Y=y](Y\langle B',B\rangle^{u}_{((R,EP))})$$

and we define

$$\mathrm{sub}[R,Y\langle B',B\rangle = y, EPu]^{*}: Y\langle B',B\rangle^{u}_{(R,EP)} \to \mathrm{iso}(R,y\langle B',B\rangle,EPu)$$

and

$$\mathrm{sub}[[R,Y\langle B',B\rangle = y, EPu]]^{*}: Y\langle B',B\rangle^{u}_{((R,EP))} \to \mathrm{iso}((R,y\langle B',B\rangle,EPu))$$

to be the R-epimorphisms induced by $\mathrm{sub}[R,Y=y]$ and we observe that

$$(1) \quad \begin{cases} \mathrm{iso}(R,Y\langle B',B\rangle,EPu) = \mathrm{iso}(R,y\langle B,B\rangle,EPu) \\ \text{and} \\ \mathrm{iso}((R,y\langle B',B\rangle,EPu)) = \mathrm{iso}((R,y\langle B,B\rangle,EPu)). \end{cases}$$

§42. Initial forms for scales and regular nets

Let E be a scale. Let R be a ring and let Y be an indeterminate net over R such that $\ell(Y) = \ell(E)$. Let $B \in [1,o(\ell(E))-1]$ and $B' \in [1,B]$. Let $u \in Q$. Let y be an R-net with $\ell(y) = \ell(E)$ such that $y\langle B\rangle$ is R-regular.

We define

$$\mathrm{info}[R,y\langle B',B\rangle=Y,E=u]: \mathrm{iso}(R,y\langle B',B\rangle,E\geq u) \to \mathrm{res}(R,y\langle B\rangle)[Y]$$

to be the $\mathrm{res}[R,y\langle B\rangle]$-homomorphism induced by

$$\mathrm{info}[R,y\langle B\rangle = Y,E(B,o(\ell(E)),o(\ell(E)),0) = u]$$

and we observe that

$$\mathrm{im}(\mathrm{info}[R,y\langle B',B\rangle = Y,E = u])$$

$$= \mathrm{Iso}(\mathrm{res}(R,y\langle B\rangle),Y\langle B,B\rangle,E = u)$$

and we define

$$\mathrm{info}[R,y\langle B',B\rangle = Y,E = u]^*: \mathrm{iso}(R,y\langle B',B\rangle,E \geq u) \downarrow \mathrm{Iso}(\mathrm{res}(R,y\langle B\rangle),Y\langle B,B\rangle,E = u)$$

to be the $\mathrm{res}[R,y\langle B\rangle]$-epimorphism induced by $\mathrm{info}[R,y\langle B',B\rangle = Y,E = u]$ and we note that the following triangle

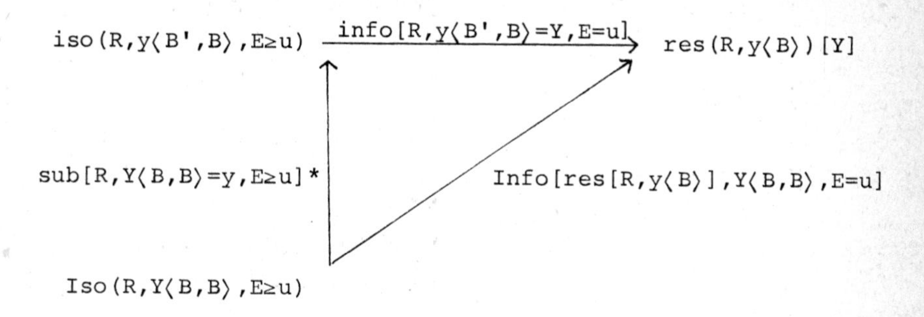

is commutative.

We define

$$\text{info}[[R,y\langle B',B\rangle=Y,E=u]]:\ \text{iso}((R,y\langle B',B\rangle,E\geq u)) \to \text{res}(R,y\langle B\rangle)[Y]$$

to be the res[R,y⟨B⟩]-homomorphism induced by

$$\text{info}[R,y\langle B\rangle=Y,E((B,o(\ell(E)),o(\ell(E)),0))=u]$$

and we observe that

$$\text{im}(\text{info}[[R,y\langle B',B\rangle=Y,E=u]])$$

$$= \text{Iso}((\text{res}(R,y\langle B\rangle),Y\langle B,B\rangle,E=u))$$

and we define

$$\text{info}[[R,y\langle B',B\rangle=Y,E=u]]^*:\ \text{iso}((R,y\langle B',B\rangle,E\geq u)) \downarrow \text{Iso}((\text{res}(R,y\langle B\rangle),Y\langle B,B\rangle,E=u))$$

to be the res[R,y⟨B⟩]-epimorphism induced by info[[R,y⟨B',B⟩=Y,E=u]] and we note that the following triangle

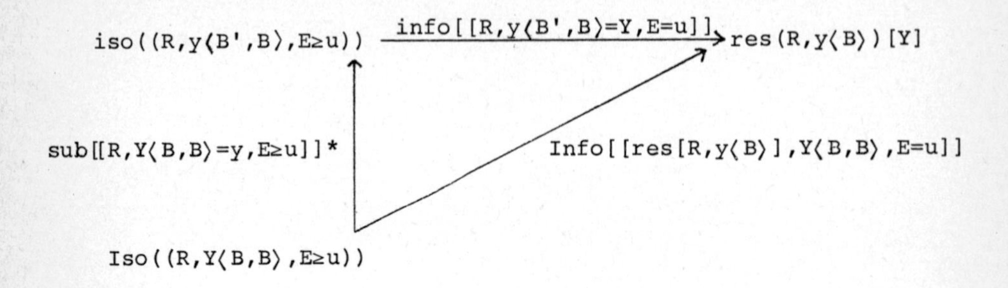

is commutative.

Given any I where

either I is an ideal in R with $y\langle B\rangle^{1}_{R} \subset I$

or I = $\bar{x}$ where $\bar{x}$ is an R-string with $y\langle B\rangle^{1}_{R} \subset \bar{x}^{1}_{R}$

or I = $\bar{x}\langle\bar{t}\rangle$ where $\bar{x}$ is an R-string and $\bar{t}$ is a string-restriction with $y\langle B\rangle^{1}_{R} \subset \bar{x}\langle\bar{t}\rangle^{1}_{R}$

or I = $\bar{y}$ where $\bar{y}$ is an R-net with $y\langle B\rangle^{1}_{R} \subset \bar{y}^{1}_{R}$

or I = $\bar{y}\langle\bar{t}\rangle$ where $\bar{y}$ is an R-net and $\bar{t}$ is a net-restriction with $y\langle B\rangle^{1}_{R} \subset \bar{y}\langle\bar{t}\rangle^{1}_{R}$

we define

$$\text{info}[(R,I),y\langle B',B\rangle=Y,E=u]: \text{iso}(R,y\langle B',B\rangle,E\geq u) \to \text{res}(R,I)[Y]$$

and

$$\text{info}[[(R,I),y\langle B',B\rangle=Y,E=u]]: \text{iso}((R,y\langle B',B\rangle,E\geq u)) \to \text{res}(R,I)[Y]$$

to be the res[R,I]-homomorphisms induced by

$$\mathrm{info}[(R,I),Y\langle B\rangle,E(B,o(\ell(E)),o(\ell(E)),0)=u]$$

and

$$\mathrm{info}[(R,I),Y\langle B\rangle,E((B,o(\ell(e)),o(\ell(E)),0))=u]$$

respectively and we observe that

$$\mathrm{im}(\mathrm{info}[(R,I),y\langle B',B\rangle=Y,E=u])$$
$$= \mathrm{Iso}(\mathrm{res}(R,I),Y\langle B,B\rangle,E=u)$$

and

$$\mathrm{im}(\mathrm{info}[[(R,I),y\langle B',B\rangle=Y,E=u]])$$
$$= \mathrm{Iso}((\mathrm{res}(R,I),Y\langle B,B\rangle,E=u))$$

and we define

$$\mathrm{info}[(R,I),y\langle B',B\rangle=Y,E=u]^*:\ \mathrm{iso}(R,y\langle B',B\rangle,E\geq u) \longrightarrow \mathrm{Iso}(\mathrm{res}(R,I),Y\langle B,B\rangle,E=u)$$

and

$$\mathrm{info}[[(R,I),y\langle B',B\rangle=Y,E=u]]^*:\ \mathrm{iso}((R,y\langle B',B\rangle,E\geq u)) \longrightarrow \mathrm{Iso}((\mathrm{res}(R,I),Y\langle B,B\rangle,E=u))$$

to be the res[R,I]-epimorphisms induced by

$$\mathrm{info}[(R,I),y\langle B',B\rangle=Y,E=u]$$

and

$$\text{info}[[(R,I),y\langle B',B\rangle=Y,E=u]]$$

respectively and we note that the following two triangles

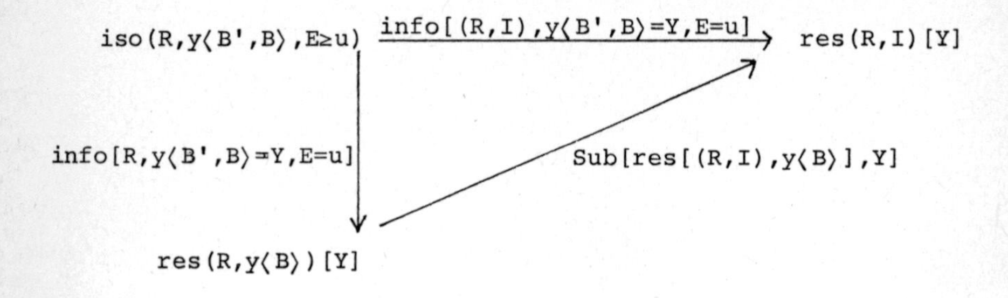

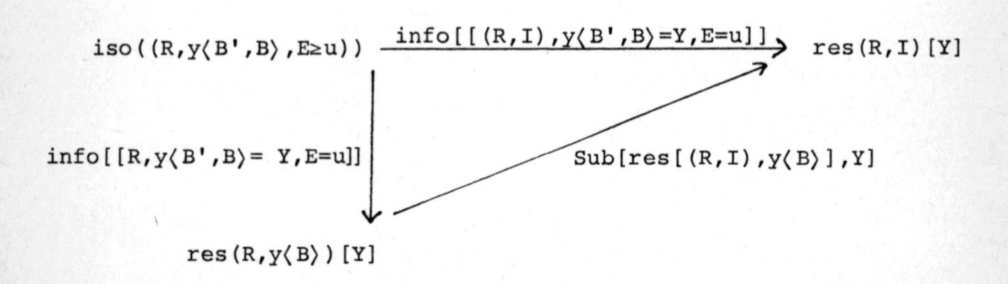

are commutative.

If $y\langle B\rangle^{1}_{R} \subset M(R)$ then we put

$$\text{info}[(R),y\langle B',B\rangle=Y,E=u] = \text{info}[(R,M(R)),y\langle B',B\rangle=Y,E=u]$$

and

$$\text{info}[(R),y\langle B',B\rangle=Y,E=u]^* = \text{info}[(R,M(R)),y\langle B',B\rangle=Y,E=u]^*$$

and we put

$$\text{info}[[(R),y\langle B',B\rangle=Y,E=u]] = \text{info}[[(R,M(R)),y\langle B',B\rangle=Y,E=u]]$$

and

$$\text{info}[[(R),y\langle B',B\rangle=Y,E=u]]^* = \text{info}[[(R,M(R)),y\langle B',B\rangle=Y,E=u]]^*.$$

§43. Isobars for protochips

Let e be a protochip. Let R be a ring and let Y be an indeterminate net over R with $\ell(Y) = \ell(e)$. Let $B \in [1,o(\ell(E))-1]$ and $B' \in [1,B]$. Let $u \in Q$ and $P \in \{=,\geq\}$. Let y be an R-net with $\ell(y) = \ell(e)$.

We define

$$\mathrm{iso}(R,y\langle B',B\rangle,ePu) = \mathrm{sub}[R,Y=y](Y\langle B',B\rangle^{u}_{(R,eP)})$$

and

$$\mathrm{iso}((R,y\langle B',B\rangle,ePu)) = \mathrm{sub}[R,Y=y](Y\langle B',B\rangle^{u}_{((R,eP))})$$

and we note that by §40 we have

$$(1)\quad \begin{cases} \mathrm{iso}(R,y\langle B',B\rangle,ePu) = \mathrm{iso}(R,y\langle B',B\rangle,e^{**}Pu) \\ \text{and} \\ \mathrm{iso}((R,y\langle B',B\rangle,ePu)) = \mathrm{iso}((R,y\langle B',B\rangle,e^{**}Pu)) \,. \end{cases}$$

§44. Initial forms for protochips and monic polynomials

Let e be a protochip. Let R be a ring and let Y be an indeterminate net over R with $\ell(Y) = \ell(e)$. Let $B \in [1, o\,\ell(e))-1]$ and $B' \in [1,B]$. Let $u \in Q$. Let y be an R-net with $\ell(y) = \ell(e)$ such that $y\langle B\rangle$ is R-regular.

We define

$$\begin{aligned}
&\mathrm{nonmon}(R, y\langle B',B\rangle, e = u) \\
&= \Big\{ f \in \mathrm{iso}(R, y\langle B',B\rangle, e \geq u) : \\
&\quad \mathrm{info}[R, y\langle B',B\rangle = Y, e^{**} = u](f) \\
&\quad \in \mathrm{Nonmon}(\mathrm{res}(R, y\langle B\rangle), Y\langle B,B\rangle, e = u) \Big\}
\end{aligned}$$

and

$$\begin{aligned}
&\mathrm{nonmon}((R, y\langle B',B\rangle, e = u)) \\
&= \Big\{ f \in \mathrm{iso}((R, y\langle B',B\rangle, e \geq u)) : \\
&\quad \mathrm{info}[[R, y\langle B',B\rangle, e^{**} = u]](f) \\
&\quad \in \mathrm{Nonmon}((\mathrm{res}(R, y\langle B\rangle), Y\langle B,B\rangle, e = u)) \Big\}
\end{aligned}$$

and we define

$$\begin{aligned}
&\mathrm{mon}(R, y\langle B',B\rangle, e = u) \\
&= \Big\{ f \in \mathrm{iso}(R, y\langle B',B\rangle, e \geq u) : \\
&\quad \mathrm{info}[R, y\langle B',B\rangle = Y, e^{**} = u](f) \\
&\quad \in \mathrm{Mon}(\mathrm{res}(R, y\langle B\rangle), Y\langle B,B\rangle, e = u) \Big\}
\end{aligned}$$

and

$$\begin{aligned}
&\text{mon}((R,y\langle B',B\rangle,e=u))\\
&=\Big\{f\in \text{iso}((R,y\langle B',B\rangle,e\geq u)):\\
&\quad \text{info}[[R,y\langle B',B\rangle = Y,e^{**}=u]](f)\\
&\quad \in \text{Mon}((\text{res}(R,y\langle B\rangle),Y\langle B,B\rangle,e=u))\Big\}\ .
\end{aligned}$$

Given any I where

$$\left\{\begin{array}{l}
\text{either } I \text{ is an ideal in } R \text{ with } y\langle B\rangle_R^1 \subset I\\
\text{or } I=\bar{x} \text{ where } \bar{x} \text{ is an } R\text{-string with } y\langle B\rangle_R^1 \subset \bar{x}_R^1\\
\text{or } I=\bar{x}\langle\bar{t}\rangle \text{ where } \bar{x} \text{ is an } R\text{-string and } \bar{t} \text{ is a string-restriction}\\
\qquad\qquad\qquad\qquad\qquad\qquad \text{with } y\langle B\rangle_R^1 \subset \bar{x}\langle\bar{t}\rangle_R^1\\
\text{or } I=\bar{y} \text{ where } \bar{y} \text{ is an } R\text{-net with } y\langle B\rangle_R^1 \subset \bar{y}_R^1\\
\text{or } I=\bar{y}\langle\bar{t}\rangle \text{ where } \bar{y} \text{ is an } R\text{-net and } \bar{t} \text{ is a net-restriction}\\
\qquad\qquad\qquad\qquad\qquad\qquad \text{with } y\langle B\rangle_R^1 \subset \bar{y}\langle\bar{t}\rangle_R^1
\end{array}\right.$$

we define

$$\begin{aligned}
&\text{nonmon}((R,I),y\langle B',B\rangle,e=u)\\
&=\Big\{f\in \text{iso}(R,y\langle B',B\rangle,e\geq u):\\
&\quad \text{info}[(R,I),y\langle B',B\rangle = Y,e^{**}=u](f)\\
&\quad \in \text{Nonmon}(\text{res}(R,I),Y\langle B,B\rangle,e=u)\Big\}
\end{aligned}$$

and

$$\begin{aligned}
&\text{nonmon}(((R,I),y\langle B',B\rangle,e=u))\\
&=\Big\{f\in \text{iso}((R,y\langle B',B\rangle,e\geq u)):\\
&\quad \text{info}[[(R,I),y\langle B',B\rangle = Y,e^{**}=u]](f)\\
&\quad \in \text{Nonmon}((\text{res}(R,I),Y\langle B,B\rangle,e=u))\Big\}
\end{aligned}$$

and we define

$$\text{mon}((R,I),y\langle B',B\rangle,e=u)$$
$$= \Big\{f \in \text{iso}(R,y\langle B',B\rangle,e \geq u): \text{info}[(R,I),y\langle B',B\rangle = Y,e^{**}=u](f) \in \text{Mon}(\text{res}(R,I),Y\langle B,B\rangle,e=u)\Big\}$$

and

$$\text{mon}(((R,I),y\langle B',B\rangle,e=u))$$
$$= \Big\{f \in \text{iso}((R,y\langle B',B\rangle,e \geq u)): \text{info}[[(R,I),y\langle B',B\rangle = Y,e^{**}=u]](f) \in \text{Mon}((\text{res}(R,I),Y\langle B,B\rangle,e=u))\Big\} .$$

If $y\langle B\rangle_R^1 \subset M(R)$ then we define

$$\text{nonmon}((R\ ),y\langle B',B\rangle,e=u)$$
$$= \text{nonmon}((R,M(R)),y\langle B',B\rangle,e=u)$$

and

$$\text{nonmon}(((R),y\langle B',B\rangle,e=u))$$
$$= \text{nonmon}(((R,M(R)),y\langle B',B\rangle,e=u))$$

and we define

$$\text{mon}((R),y\langle B',B\rangle,e=u)$$
$$= \text{mon}((R,M(R)),y\langle B',B\rangle,e=u)$$

and

$$\mathrm{mon}(((R), y\langle B', B\rangle, e = u))$$

$$= \mathrm{mon}(((R, M(R)), y\langle B', B\rangle, e = u)).$$

Index of definitions

Index of notations

Vol. 759: R. L. Epstein, Degrees of Unsolvability: Structure and Theory. XIV, 216 pages. 1979.

Vol. 760: H.-O. Georgii, Canonical Gibbs Measures. VIII, 190 pages. 1979.

Vol. 761: K. Johannson, Homotopy Equivalences of 3-Manifolds with Boundaries. 2, 303 pages. 1979.

Vol. 762: D. H. Sattinger, Group Theoretic Methods in Bifurcation Theory. V, 241 pages. 1979.

Vol. 763: Algebraic Topology, Aarhus 1978. Proceedings, 1978. Edited by J. L. Dupont and H. Madsen. VI, 695 pages. 1979.

Vol. 764: B. Srinivasan, Representations of Finite Chevalley Groups. XI, 177 pages. 1979.

Vol. 765: Padé Approximation and its Applications. Proceedings, 1979. Edited by L. Wuytack. VI, 392 pages. 1979.

Vol. 766: T. tom Dieck, Transformation Groups and Representation Theory. VIII, 309 pages. 1979.

Vol. 767: M. Namba, Families of Meromorphic Functions on Compact Riemann Surfaces. XII, 284 pages. 1979.

Vol. 768: R. S. Doran and J. Wichmann, Approximate Identities and Factorization in Banach Modules. X, 305 pages. 1979.

Vol. 769: J. Flum, M. Ziegler, Topological Model Theory. X, 151 pages. 1980.

Vol. 770: Séminaire Bourbaki vol. 1978/79 Exposés 525–542. IV, 341 pages. 1980.

Vol. 771: Approximation Methods for Navier-Stokes Problems. Proceedings, 1979. Edited by R. Rautmann. XVI, 581 pages. 1980.

Vol. 772: J. P. Levine, Algebraic Structure of Knot Modules. XI, 104 pages. 1980.

Vol. 773: Numerical Analysis. Proceedings, 1979. Edited by G. A. Watson. X, 184 pages. 1980.

Vol. 774: R. Azencott, Y. Guivarc'h, R. F. Gundy, Ecole d'Eté de Probabilités de Saint-Flour VIII-1978. Edited by P. L. Hennequin. XIII, 334 pages. 1980.

Vol. 775: Geometric Methods in Mathematical Physics. Proceedings, 1979. Edited by G. Kaiser and J. E. Marsden. VII, 257 pages. 1980.

Vol. 776: B. Gross, Arithmetic on Elliptic Curves with Complex Multiplication. V, 95 pages. 1980.

Vol. 777: Séminaire sur les Singularités des Surfaces. Proceedings, 1976-1977. Edited by M. Demazure, H. Pinkham and B. Teissier. IX, 339 pages. 1980.

Vol. 778: SK1 von Schiefkörpern. Proceedings, 1976. Edited by P. Draxl and M. Kneser. II, 124 pages. 1980.

Vol. 779: Euclidean Harmonic Analysis. Proceedings, 1979. Edited by J. J. Benedetto. III, 177 pages. 1980.

Vol. 780: L. Schwartz, Semi-Martingales sur des Variétés, et Martingales Conformes sur des Variétés Analytiques Complexes. XV, 132 pages. 1980.

Vol. 781: Harmonic Analysis Iraklion 1978. Proceedings 1978. Edited by N. Petridis, S. K. Pichorides and N. Varopoulos. V, 213 pages. 1980.

Vol. 782: Bifurcation and Nonlinear Eigenvalue Problems. Proceedings, 1978. Edited by C. Bardos, J. M. Lasry and M. Schatzman. VIII, 296 pages. 1980.

Vol. 783: A. Dinghas, Wertverteilung meromorpher Funktionen in ein- und mehrfach zusammenhängenden Gebieten. Edited by R. Nevanlinna and C. Andreian Cazacu. XIII, 145 pages. 1980.

Vol. 784: Séminaire de Probabilités XIV. Proceedings, 1978/79. Edited by J. Azéma and M. Yor. VIII, 546 pages. 1980.

Vol. 785: W. M. Schmidt, Diophantine Approximation. X, 299 pages. 1980.

Vol. 786: I. J. Maddox, Infinite Matrices of Operators. V, 122 pages. 1980.

Vol. 787: Potential Theory, Copenhagen 1979. Proceedings, 1979. Edited by C. Berg, G. Forst and B. Fuglede. VIII, 319 pages. 1980.

Vol. 788: Topology Symposium, Siegen 1979. Proceedings, 1979. Edited by U. Koschorke and W. D. Neumann. VIII, 495 pages. 1980.

Vol. 789: J. E. Humphreys, Arithmetic Groups. VII, 158 pages. 1980.

Vol. 790: W. Dicks, Groups, Trees and Projective Modules. IX, 127 pages. 1980.

Vol. 791: K. W. Bauer and S. Ruscheweyh, Differential Operators for Partial Differential Equations and Function Theoretic Applications. V, 258 pages. 1980.

Vol. 792: Geometry and Differential Geometry. Proceedings, 1979. Edited by R. Artzy and I. Vaisman. VI, 443 pages. 1980.

Vol. 793: J. Renault, A Groupoid Approach to C*-Algebras. III, 160 pages. 1980.

Vol. 794: Measure Theory, Oberwolfach 1979. Proceedings 1979. Edited by D. Kölzow. XV, 573 pages. 1980.

Vol. 795: Séminaire d'Algèbre Paul Dubreil et Marie-Paule Malliavin. Proceedings 1979. Edited by M. P. Malliavin. V, 433 pages. 1980.

Vol. 796: C. Constantinescu, Duality in Measure Theory. IV, 197 pages. 1980.

Vol. 797: S. Mäki, The Determination of Units in Real Cyclic Sextic Fields. III, 198 pages. 1980.

Vol. 798: Analytic Functions, Kozubnik 1979. Proceedings. Edited by J. Ławrynowicz. X, 476 pages. 1980.

Vol. 799: Functional Differential Equations and Bifurcation. Proceedings 1979. Edited by A. F. Izé. XXII, 409 pages. 1980.

Vol. 800: M.-F. Vignéras, Arithmétique des Algèbres de Quaternions. VII, 169 pages. 1980.

Vol. 801: K. Floret, Weakly Compact Sets. VII, 123 pages. 1980.

Vol. 802: J. Bair, R. Fourneau, Etude Géometrique des Espaces Vectoriels II. VII, 283 pages. 1980.

Vol. 803: F.-Y. Maeda, Dirichlet Integrals on Harmonic Spaces. X, 180 pages. 1980.

Vol. 804: M. Matsuda, First Order Algebraic Differential Equations. VII, 111 pages. 1980.

Vol. 805: O. Kowalski, Generalized Symmetric Spaces. XII, 187 pages. 1980.

Vol. 806: Burnside Groups. Proceedings, 1977. Edited by J. L. Mennicke. V, 274 pages. 1980.

Vol. 807: Fonctions de Plusieurs Variables Complexes IV. Proceedings, 1979. Edited by F. Norguet. IX, 198 pages. 1980.

Vol. 808: G. Maury et J. Raynaud, Ordres Maximaux au Sens de K. Asano. VIII, 192 pages. 1980.

Vol. 809: I. Gumowski and Ch. Mira, Recurences and Discrete Dynamic Systems. VI, 272 pages. 1980.

Vol. 810: Geometrical Approaches to Differential Equations. Proceedings 1979. Edited by R. Martini. VII, 339 pages. 1980.

Vol. 811: D. Normann, Recursion on the Countable Functionals. VIII, 191 pages. 1980.

Vol. 812: Y. Namikawa, Toroidal Compactification of Siegel Spaces. VIII, 162 pages. 1980.

Vol. 813: A. Campillo, Algebroid Curves in Positive Characteristic. V, 168 pages. 1980.

Vol. 814: Séminaire de Théorie du Potentiel, Paris, No. 5. Proceedings. Edited by F. Hirsch et G. Mokobodzki. IV, 239 pages. 1980.

Vol. 815: P. J. Slodowy, Simple Singularities and Simple Algebraic Groups. XI, 175 pages. 1980.

Vol. 816: L. Stoica, Local Operators and Markov Processes. VIII, 104 pages. 1980.

Vol. 817: L. Gerritzen, M. van der Put, Schottky Groups and Mumford Curves. VIII, 317 pages. 1980.

Vol. 818: S. Montgomery, Fixed Rings of Finite Automorphism Groups of Associative Rings. VII, 126 pages. 1980.

Vol. 819: Global Theory of Dynamical Systems. Proceedings, 1979. Edited by Z. Nitecki and C. Robinson. IX, 499 pages. 1980.

Vol. 820: W. Abikoff, The Real Analytic Theory of Teichmüller Space. VII, 144 pages. 1980.

Vol. 821: Statistique non Paramétrique Asymptotique. Proceedings, 1979. Edited by J.-P. Raoult. VII, 175 pages. 1980.

Vol. 822: Séminaire Pierre Lelong–Henri Skoda, (Analyse) Années 1978/79. Proceedings. Edited by P. Lelong et H. Skoda. VIII, 356 pages, 1980.

Vol. 823: J. Král, Integral Operators in Potential Theory. III, 171 pages. 1980.

Vol. 824: D. Frank Hsu, Cyclic Neofields and Combinatorial Designs. VI, 230 pages. 1980.

Vol. 825: Ring Theory, Antwerp 1980. Proceedings. Edited by F. van Oystaeyen. VII, 209 pages. 1980.

Vol. 826: Ph. G. Ciarlet et P. Rabier, Les Equations de von Kármán. VI, 181 pages. 1980.

Vol. 827: Ordinary and Partial Differential Equations. Proceedings, 1978. Edited by W. N. Everitt. XVI, 271 pages. 1980.

Vol. 828: Probability Theory on Vector Spaces II. Proceedings, 1979. Edited by A. Weron. XIII, 324 pages. 1980.

Vol. 829: Combinatorial Mathematics VII. Proceedings, 1979. Edited by R. W. Robinson et al.. X, 256 pages. 1980.

Vol. 830: J. A. Green, Polynomial Representations of GL_n. VI, 118 pages. 1980.

Vol. 831: Representation Theory I. Proceedings, 1979. Edited by V. Dlab and P. Gabriel. XIV, 373 pages. 1980.

Vol. 832: Representation Theory II. Proceedings, 1979. Edited by V. Dlab and P. Gabriel. XIV, 673 pages. 1980.

Vol. 833: Th. Jeulin, Semi-Martingales et Grossissement d'une Filtration. IX, 142 Seiten. 1980.

Vol. 834: Model Theory of Algebra and Arithmetic. Proceedings, 1979. Edited by L. Pacholski, J. Wierzejewski, and A. J. Wilkie. VI, 410 pages. 1980.

Vol. 835: H. Zieschang, E. Vogt and H.-D. Coldewey, Surfaces and Planar Discontinuous Groups. X, 334 pages. 1980.

Vol. 836: Differential Geometrical Methods in Mathematical Physics. Proceedings, 1979. Edited by P. L. García, A. Pérez-Rendón, and J. M. Souriau. XII, 538 pages. 1980.

Vol. 837: J. Meixner, F. W. Schäfke and G. Wolf, Mathieu Functions and Spheroidal Functions and their Mathematical Foundations Further Studies. VII, 126 pages. 1980.

Vol. 838: Global Differential Geometry and Global Analysis. Proceedings 1979. Edited by D. Ferus et al. XI, 299 pages. 1981.

Vol. 839: Cabal Seminar 77 – 79. Proceedings. Edited by A. S. Kechris, D. A. Martin and Y. N. Moschovakis. V, 274 pages. 1981.

Vol. 840: D. Henry, Geometric Theory of Semilinear Parabolic Equations. IV, 348 pages. 1981.

Vol. 841: A. Haraux, Nonlinear Evolution Equations- Global Behaviour of Solutions. XII, 313 pages. 1981.

Vol. 842: Séminaire Bourbaki vol. 1979/80. Exposés 543–560. IV, 317 pages. 1981.

Vol. 843: Functional Analysis, Holomorphy, and Approximation Theory. Proceedings. Edited by S. Machado. VI, 636 pages. 1981.

Vol. 844: Groupe de Brauer. Proceedings. Edited by M. Kervaire and M. Ojanguren. VII, 274 pages. 1981.

Vol. 845: A. Tannenbaum, Invariance and System Theory: A
and Geometric Aspects. X, 161 pages. 1981.

Vol. 846: Ordinary and Partial Differential Equations, Proc
Edited by W. N. Everitt and B. D. Sleeman. XIV, 384 pag

Vol. 847: U. Koschorke, Vector Fields and Other Vecto
Morphisms – A Singularity Approach. IV, 304 pages. 1981.

Vol. 848: Algebra, Carbondale 1980. Proceedings. Ed.
Amayo. VI, 298 pages. 1981.

Vol. 849: P. Major, Multiple Wiener-Itô Integrals. VII, 127 pag

Vol. 850: Séminaire de Probabilités XV. 1979/80. Avec table
des exposés de 1966/67 à 1978/79. Edited by J. Azéma an
IV, 704 pages. 1981.

Vol. 851: Stochastic Integrals. Proceedings, 1980. Edite
Williams. IX, 540 pages. 1981.

Vol. 852: L. Schwartz, Geometry and Probability in Banach
X, 101 pages. 1981.

Vol. 853: N. Boboc, G. Bucur, A. Cornea, Order and Co
Potential Theory: H-Cones. IV, 286 pages. 1981.

Vol. 854: Algebraic K-Theory. Evanston 1980. Proceeding
by E. M. Friedlander and M. R. Stein. V, 517 pages. 1981.

Vol. 855: Semigroups. Proceedings 1978. Edited by H. Jü
M. Petrich and H. J. Weinert. V, 221 pages. 1981.

Vol. 856: R. Lascar, Propagation des Singularités des S
d'Equations Pseudo-Différentielles à Caractéristiques de
cités Variables. VIII, 237 pages. 1981.

Vol. 857: M. Miyanishi. Non-complete Algebraic Surface
244 pages. 1981.

Vol. 858: E. A. Coddington, H. S. V. de Snoo: Regular Bounda
Problems Associated with Pairs of Ordinary Differential Expr
V, 225 pages. 1981.

Vol. 859: Logic Year 1979–80. Proceedings. Edited by M.
J. Schmerl and R. Soare. VIII, 326 pages. 1981.

Vol. 860: Probability in Banach Spaces III. Proceedings, 198
by A. Beck. VI, 329 pages. 1981.

Vol. 861: Analytical Methods in Probability Theory. Proceedin
Edited by D. Dugué, E. Lukacs, V. K. Rohatgi. X, 183 pag

Vol. 862: Algebraic Geometry. Proceedings 1980. Edited b
gober and P. Wagreich. V, 281 pages. 1981.

Vol. 863: Processus Aléatoires à Deux Indices. Proceeding
Edited by H. Korezlioglu, G. Mazziotto and J. Szpirglas. V, 27
1981.

Vol. 864: Complex Analysis and Spectral Theory. Proc
1979/80. Edited by V. P. Havin and N. K. Nikol'skii, VI, 48
1981.

Vol. 865: R. W. Bruggeman, Fourier Coefficients of Auto
Forms. III, 201 pages. 1981.

Vol. 866: J.-M. Bismut, Mécanique Aléatoire. XVI, 563 pag

Vol. 867: Séminaire d'Algèbre Paul Dubreil et Marie-Paule M
Proceedings, 1980. Edited by M.-P. Malliavin. V, 476 pag

Vol. 868: Surfaces Algébriques. Proceedings 1976–78. E
J. Giraud, L. Illusie et M. Raynaud. V, 314 pages. 1981.

Vol. 869: A. V. Zelevinsky, Representations of Finite Classica
IV, 184 pages. 1981.

Vol. 870: Shape Theory and Geometric Topology. Proceedin
Edited by S. Mardešić and J. Segal. V, 265 pages. 1981.

Vol. 871: Continuous Lattices. Proceedings, 1979. Edited by
schewski and R.-E. Hoffmann. X, 413 pages. 1981.

Vol. 872: Set Theory and Model Theory. Proceedings, 197
by R. B. Jensen and A. Prestel. V, 174 pages. 1981.